Today Is a Great Day!

Today Is a Great Day! New Attitudes for Attaining Project Success is a beacon of positivity and inspiration to project managers who, in their day-to-day work, are beset with challenges and uncertainty. Through a blend of personal anecdotes, insightful reflections, and practical wisdom, the book shows project managers how to embrace each day with a sense of optimism and purpose.

At the book's core is the message that attitude shapes outcomes. This powerful message helps readers to cultivate a mindset of gratitude and resilience, regardless of the circumstances they may face. Drawing from personal experience navigating the ups and downs of delivering projects, Bucero explains how adopting a positive outlook can transform obstacles into opportunities and setbacks into steppingstones.

This call to action encourages readers to apply the insights gained from the book to their own projects. It guides project managers through the steps toward a positive attitude that fosters a project team culture focused on growth and project success. The book:

- Gives some examples of "positive attitude – project success" for project and organizational success
- Helps project managers and executives create a positive atmosphere to manage projects successfully
- Shows how to understand and empathize with all project stakeholders to work efficiently together

This book is an attitude implementation guide filled with tools, real-world examples, and global case studies that address an international audience. Based on the author's award-winning background as a project and program manager, as well as a project management trainer and consultant, the book shares case studies, best practices, and mindsets, as well as exercises and checklists, to help project managers and executives adopt winning attitudes that can promote project success.

Today Is a Great Day!
New Attitudes for Attaining Project Success

Alfonso Bucero

CRC Press
Taylor & Francis Group
Boca Raton London New York

CRC Press is an imprint of the
Taylor & Francis Group, an **informa** business

AN AUERBACH BOOK

Cover image: Shutterstock

First edition published 2025
2385 NW Executive Center Drive, Suite 320, Boca Raton FL 33431

and by CRC Press
4 Park Square, Milton Park, Abingdon, Oxon, OX14 4RN

CRC Press is an imprint of Taylor & Francis Group, LLC

ISBN: 978-1-032-77552-4 (hbk)
ISBN: 978-1-032-77975-1 (pbk)
ISBN: 978-1-003-48567-4 (ebk)

DOI: 10.1201/9781003485674

Typeset in Minion
by SPi Technologies India Pvt Ltd (Straive)

Contents

Foreword

One of the best decisions I made in my career was inviting Alfonso Bucero to present at an internal conference where we both worked. His passion, persistence, and patience became evident in numerous interactions over many years, leading to collaboration on three books with subsequent editions, and presenting together at conferences and training sessions—all encompassing a labor of love.

When I first saw Alfonso state "today is a good day," I wondered if something marvelous happened on that day in his life. Then I discovered that it is his recurring motto or slogan. He constantly repeats—and lives—this statement every day, which reminds and frames his attitude toward life. He goes on to say, by embracing this attitude, "tomorrow will be even better."

As readers, we are now fortunate to learn from his experiences that he shares so eloquently *Today Is a Great Day! New Attitudes for Attaining Project Success* builds upon his attitude and approach to project management. This new book features numerous tools, such as self-assessments, that help readers adopt, adapt, and apply the concepts. He shares mind maps which help illustrate key topics and approaches, especially valuable for visual learners. His case studies provide rich testimonials to the validity and application of the positive attitude that he espouses.

Not only is Alfonso an experienced practitioner, consultant, and educator, but he also added depth by his research and learnings in achieving a Ph.D. He is not immune from difficulties in his life; he provides a positive mindset in learning and becoming a better, inspirational person in how he reflects upon various challenges.

Humor and fun in the workplace are other elements that Alfonso brings to his writings and personal interactions. He is a master story and joke teller.

I treasure the friendship and knowledge shared with Alfonso over many years. This book is a masterpiece that reflects the genuine person he is. We are blessed by his willingness to share his wisdom. The result is a necessary and important book.

As you read this book and apply its principles or have an opportunity to see him speak in public, you will be inspired to broaden, deepen, and achieve more in life and in the projects you manage.

Randall L. Englund, BSEE, MBA
Executive Consultant, Englund Project Management Consultancy
Co-author of *Creating an Environment for Successful Projects, The Complete Project Manager, The Complete Project Manager's Toolkit*, and *Project Sponsorship.*

Acknowledgments

By May 2000, I had the opportunity to visit New Delhi in India, attending a Project Management Conference as a keynote speaker. It was a real challenge because my English was not very good then. However, I applied my courage to attend that conference and communicate not only through my words but also from my heart. I learned a lot at that conference about the incredible power of a positive attitude. Since then, I have started applying my three Ps (Passion, Persistence, and Patience) as my personal and professional principles. It is still working for me.

By May 2006, I had attended a PMI Leadership Program as part of the Leadership Master Class. As soon as I graduated, I committed to writing my *Today Is a Good Day* book, telling my experiences and stories about the value of a positive attitude and the difference maker for the project manager. Since 2006, I have dedicated time daily to gathering information and describing my life experiences as a project manager and human being. However, my *Today Is a Good Day* book was not published till 2010 because I had difficulty finding a book editor. Now, fourteen years later, I can test that my assumptions about using a positive attitude as a project manager are accurate, and thanks to them, you are now reading this new book.

This book addresses different sets of project stakeholders:

- All project managers and executives who contributed to my book with their opinions, experiences, and practices for this project, directly or indirectly.
- My best friend and better project professional, Randall L. Englund, always gave me great advice about this book's content, shared many metaphors that inspired my passion for project management, and encouraged me to continue writing.
- My wife Rose, my three children (Luis, Cristina, and Alfonso), who are now adults, and Manuel (my grandson) gave me unwavering support and encouragement while writing this book.
- The memory of my parents, Emilia and Alfonso, who taught me how much you need other people and invested a lot of effort and money in my development.

- The memory of Dr. David I. Cleland, who read my manuscript and gave me great ideas and suggestions when writing my previous book.
- The memory of Dr. Ginger Levin, who always encouraged me to move forward to achieve my goals.
- The memory of Dr. Davidson Frame, a good friend and Mentor who always encouraged me to grow professionally.

Thank you very much to all of them because they helped and encouraged me to finish my "TODAY IS A GREAT DAY" project.

About the Author

Alfonso Bucero, Ph.D., PMP, PMI-RMP, PfMP, PMI Fellow, and Certified Public Speaker is an independent project management consultant, author, and speaker. He was the founder, partner, and director of BUCERO PM Consulting in Spain. Bucero has a M.S. in Computer Science Engineering and a Ph.D. in Project Management. He is the author of eleven project management books, five of them co-authored with Randall L. Englund. Alfonso manages projects internationally. He delivers workshops, keynote speeches, and consults organizations on project, program, and portfolio management. His motto is Passion, Persistence, and Patience, and for him every day is a great day.

1

Your Attitude

If you don't like something, change it. If you can't change it, change your attitude.

Don't complain.

–Maya Angelou, U.S. author and poet (1928)

INTRODUCTION

I had a bad attitude about my work and the projects I oversaw early in my career. That pessimistic attitude produced more issues than benefits. In front of my superiors, team members, and coworkers, I cultivated a wrong impression of myself, not being focused on some things I did well. The outcome wasn't favorable. I spread negativity to my teammates and managers. My process of maturation caused me to shift the way I thought. I required a mindset adjustment! I transformed my world by altering my attitude. I feel obligated to share this profound and life-altering experience with you, my readers.

A DEFINITION OF ATTITUDE

The dictionary defines attitude as a body position or manner of carrying oneself, a state of mind or a feeling, disposition, or *an arrogant or hostile state of mind or personality. According to the Merriam-Webster dictionary,*

DOI: 10.1201/9781003485674-1

attitude is a bodily state of readiness to respond characteristically to a stimulus (such as an object, concept, or situation). Attitude is the preference of an individual or organization toward or away from things, events, or people. It is the spirit and perspective from which an individual, group, or organization approaches community development. Your attitude shapes all your decisions and actions. Attitude is uncompromising to define precisely, as it consists of non-tangible qualities and beliefs. We are used to talking about the attitudes of individuals, but it is essential to recognize that project teams and organizations also have attitudes. Usually, however, when discussing an organization's attitude, we use the term "organizational culture." When discussing a project's attitude, we use the term project culture. The project manager's attitude dramatically affects the team's attitude.

For instance, a vital team attitude is confidence. The development of a project presents tremendous challenges to a project team. Sometimes, it can even feel like an act of faith. Much detail is collected, analyzed, organized, and assimilated into a functional "whole." On extensive efforts, only a few key individuals may possess the total "big picture," which may be at varying levels of completeness. This ambiguity can, from time to time, test the confidence of the project team members. Given these uncertainties, how does a team feel assured and confident of success throughout the process, and have this reflected in individual team member attitudes?

Figure 1.1 shows essential qualities and beliefs that, from my experience, determine whether or not an individual, team, or organization has the attitude needed to lead or actively participate in a project successfully.

1. Respect: Every project manager must respect individuals, their quirks, culture, and reactions.
2. Responsibility: Project management commitment is critical for the success of every project. Practice your authenticity and integrity, say what you believe, and act on what you say (Englund & Graham, 2019 Englund, R., & Graham, R. J. (2019). *Creating an environment for successful projects.* Berrett-Koehler Publishers.).
3. Empathy: Please test somebody else's boots to feel how they feel. Understand where others are coming from.
4. Openness: You can be open to new ideas, alternatives, and solutions.
5. 3Ps: Cultivate your passion (enthusiasm), persistence, and patience.
6. CII: Cultivate your creativity, innovation, and intuition. You are an excellent professional, smart enough to create new solutions and innovate, and you can use your intuition based on your practical experience.

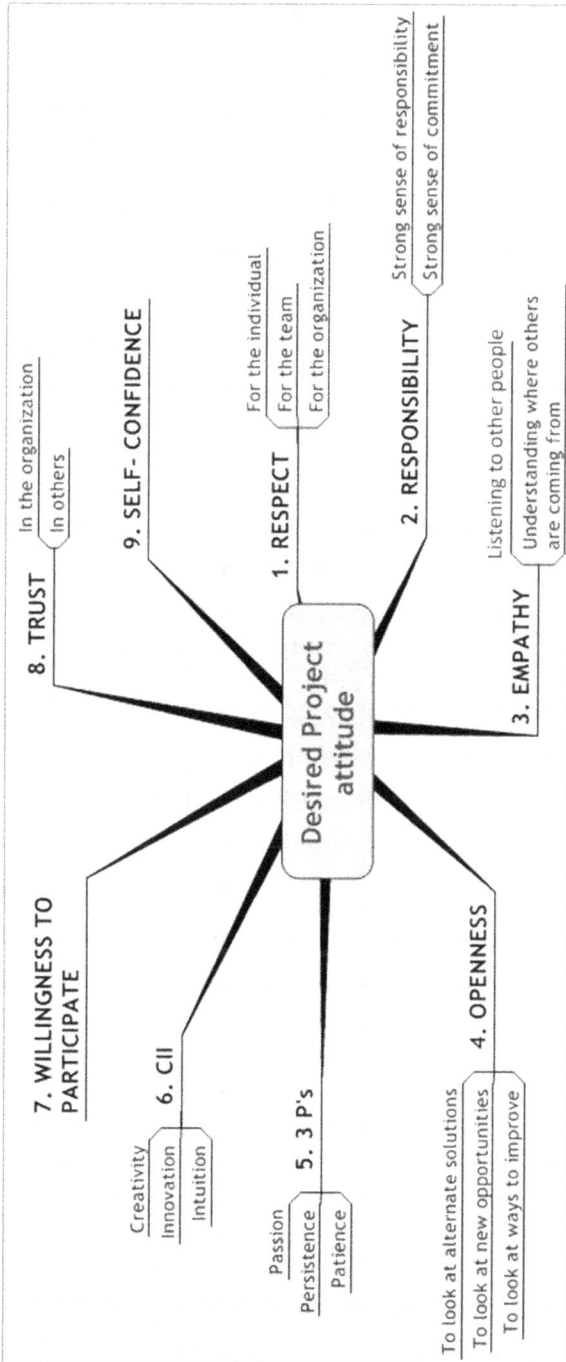

FIGURE 1.1
Desired project attitude.

7. Willingness to participate: Be always ready to contribute and force yourself to be slightly uncomfortable for personal and professional growth. Preparation is the key.
8. Trust: Trust your organization; trust your peers and colleagues. If you fail, you are strong enough to overcome any obstacle. Always look at the positive side of life.
9. Self-confidence: Be confident in yourself. You do many things well, award yourself because of that, and try to improve. You can do it because you are an excellent professional.

William James said: "*The greatest discovery of my generation is that human beings can alter their lives by altering their attitude of mind.*" I firmly believe attitude is a choice. The average project manager wants to wait for someone else to motivate them. The project manager perceives that their circumstances are responsible for how they think. But which comes first, the attitude or the circumstances? The truth is that it does not matter which comes first. No matter what happened to you yesterday in your project or organization, your attitude is your choice today. Your attitude determines your actions. Attitudes are a secret power working twenty-four hours daily, for good or bad. Attitude is a brain filter through which you experience the world. Some people see the world through a mind of optimism, while others see life through a mind of pessimism. I found some people in the middle—not very optimistic but also not very pessimistic.

Here is an example to explain the difference between a positive attitude and a negative attitude regarding a project: Imagine a manager asking a project manager, *how is your project going?* The project manager answers: *It is going wrong, as always.* With that approach, your enemies will be happy, and your friends will be sad. A better answer would be: *We are progressing well. Some project activities were delayed, and there were some project issues, but my team is taking corrective actions, and everything will be under control very soon. I'll keep you informed about the project's progress.*

People with positive attitudes focus on project solutions. People with negative attitudes concentrate on problems and issues. Project managers with negative attitudes dramatically affect project success. The attitude of the project manager to the project and the team will determine the project's attitude to the project manager. We shape our projects. We have the choice of choosing the attitude that will make a success of our projects.

Attitude is an excellent reflection of your spirit. Look at yourself. Are you happy as a project manager? Be honest. The environment you find around yourself is a mirror of your attitude. If you have a problem, then you should start asking questions. Your team will change when you change for them. I advise treating those around you as you want them to treat you. Everyone needs recognition, gratitude, and a kind word. My experience is that the attitude you start has a marked influence on the outcome of any venture. Good attitudes are often the introduction to an opportunity and the final arbiter to success. A fundamental desire of professionals is to be respected and appreciated. How can we make that happen? Instead of thinking about what's missing, count your blessings. Don't see project limitations; identify risks and see opportunities. I have managed many projects outside my city, staying away from Monday to Friday, far from home. I needed to be positive with my people despite my personal strain. I always believe I should lead by example. When I talk to project management audiences about project manager attitude, I use pictures, jokes, and video clips to help people understand and remember what I said. I demonstrate that I care about communicating effectively with the audience and will use various means to make my message as straightforward as possible. Make your attitude toward project stakeholders "fresh"—open your "project window" to get fresh air and keep people moving forward through energy.

THE CONCEPT OF PROJECT WINDOW

Your attitude is your window to project success. We all start in life with a good attitude. Just watch our children. They are always laughing and giggling. They have a cheerful disposition, and they love to explore new things. When my youngest child learned to ride a bicycle, he fell down many times but always laughed. He fell and tried again and again. He spent some days learning to ride the bike and had a lot of fun doing that activity. I observed how he never complained; his objective was to learn (Figure 1.2).

Most activities we do in a project fail in the beginning. We must try until we get it; we must be *persistent*. We experience a lot of criticism in project environments. Our project window gets splattered by nasty comments and feedback from colleagues, executives, and sometimes project sponsors. *"You cannot always control your project issues, but you can manage your*

FIGURE 1.2
Project window.

thoughts." The problem is the dirt keeps building up, and too many people do nothing about it. I had a dirty window when I worked for a multinational company as a project manager for many years. And the longer I stayed in that job, the filthier my window got. I saw no possibility of moving forward; I needed fresh air. How could I? I found a lot of negativities. Management feedback was too negative, or at least that was my perception. There was a poor project management attitude in that organization. When I asked for input from other colleagues regarding my perception, they reinforced my opinion. That encouraged me to move forward.

WASH YOUR PROJECT WINDOW

Fortunately, I discovered that all I had to do was clean off my project window and look outside. I had to improve my attitude to see the professional world again. After removing the grime from my project window, a new world opened for me. My frustration lifted. I had more confidence. For the first time in many years, I could see the magnificent possibilities my profession, as a project management practitioner, had to offer. I was much more conscious about my lack of knowledge. The opportunities for learning in the project management field became huge. Then, I could make a career move and do work I love. Now, *"Today is a great day"* is my favorite sentence. I founded my company several years ago. I passed through problems, difficulties, and issues, but now I feel free as a bird and happy doing what I want to do. I started my "project" with a positive attitude, not forgetting that no project is without problems.

We must deal with problems and issues in our lives and the projects we manage. My attitude has changed completely, and this reflects positively on my professional and personal life. It affects the projects I work on and my project stakeholders and has a significant positive business impact. Furthermore, it affects my family, my most critical project. I am happy and try to transmit positivity to everyone most of the time. Not every day can I achieve 100%, but I try to learn from my experience day by day. I believe my work is enjoyable, and I try to make it enjoyable for my team daily.

TAKE CHARGE OF YOUR ATTITUDE

Now, your job will be to keep your project window clean. My objective is to provide a little encouragement through this book. And other people can encourage you, too. But it would help if you did it; nobody else can. You always have a choice. You can leave the filth on your window and look at your projects through a smeared glass. But there are consequences to that approach, and they are not very pretty. You will go through projects with a negative attitude and feel frustrated. You will be unhappy. You will achieve only a fraction of what you are capable of achieving. There is a better way. When you remove your squeegee and clean your window, life will be brighter and sunnier. You will be healthier and happier. You will set some ambitious goals and begin to achieve them. Your dreams will come alive again. You probably think that's easy to say but challenging to do. Granted, some devastating things may have happened to you. You may have endured much suffering. Perhaps you are going through some tough times right now. But, even under the worst circumstances, you can choose your attitude. I am not saying it is easy. But I believe the choice is yours.

Let me share my personal experience managing a project: *Some years ago, I managed a project in the South of Spain. At the beginning of the project, my team was not there. I was alone before the customer because my squad was finishing another project. For almost one month, I was alone with my customer and showed a lot of enthusiasm regarding the project. I worked with my customer on the strategic part of the project and spent time analyzing various project stakeholders and talking about the commitment needed from everyone for project success. I organized presentations and workshops to create teamwork until my project team was on the customer site. I felt so stressed by that situation, but I never showed my stress in front of the customer, trying to demonstrate self-control. I exercised for one hour daily and wanted to keep myself healthy. My self-discipline was critical in that situation because the project environment is often unhealthy.* It was a very stressful situation. A happy project manager is not a professional in a particular set of circumstances but a competent professional with a set of attitudes.

ATTITUDE AND PROJECT SUCCESS

Let's say you clean off your window and develop a positive attitude. You are smiling. You sit home and think positive thoughts. Will that alone lead you to outrageous success and realizing your fondest dreams? No, it won't because there's more to success than having a great attitude. To maximize your potential and achieve your goals, I advise applying certain principles of success that have helped me achieve excellent results. I hope they will serve you in the same way. In the following chapters of this book, I take you through these principles step by step. You will learn about confronting your project fears, overcoming adversity, harnessing the power of commitment, and more. Still, you may be wondering what these success principles have to do with attitude. TODAY IS A GREAT DAY. Without a positive attitude, you cannot activate the other principles. However, the promise that attitude is everything is hollow. If you believe that attitude is everything, it may hurt you more than help you. There is another factor that stands in the way: talent. Attitude is the difference maker. Attitude is not everything, but it is one thing that can make a difference in your projects when talent is present. This book aims to show you that attitude is the difference maker in your projects. Then, I always apply two rules every day:

- Rule #1: Today is a Great Day!
- Rule #2: If today is not a Great Day for you, please apply Rule #1.

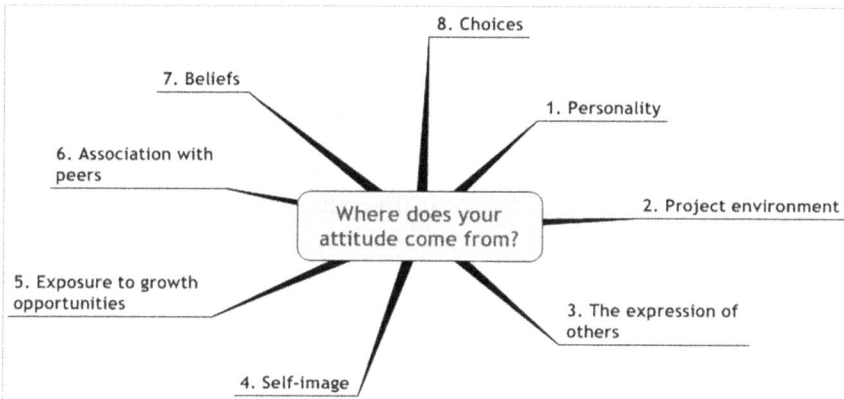

FIGURE 1.3
Where does your attitude come from?

There is not a single part of your current life that is not affected by your attitude. The attitude you carry forward from today will influence your future. You become unstoppable when you combine a positive attitude with the other success principles. If attitude is so important, you may ask yourself, where does attitude come from? See Figure 1.3.

1. Personality: Who are you? Your personality type impacts your attitude. That is not to say your personality traps you because you are not. But your attitude is affected by it.
2. Project environment: What's around you? The environment you are exposed to in your projects impacts your attitude. Do you have enough management support from your upper manager in your project? Do you have a sponsor assigned for your project? Do you have any project constraints, or deadlines? It may be hard to predict exactly what will happen to a person's attitude based on his or her early environment, but you can be certain that it made an impact of some kind. Clearly, the environment makes a difference.
3. The expression of others: What you feel. Most people can remember the harsh words of a parent, coach, or teacher even years or decades after the fact. Many times, the hurts that cause people to overreact to others come as the result of negative words from others. Likewise, positive words can have an impact on a person's attitude.
4. Self-image: How you see yourself has a tremendous impact on your attitude. Poor self-image and poor attitudes often walk hand in hand. It is hard to see anything in the world as positive if you see

yourself as negative. If you do not change your inward feelings about yourself, you will be unable to change your outward actions toward others.

5. Exposure to growth opportunities: Players must accept the cards dealt to them. However, once they have those cards in hand, they can choose how they will play them. They decide what risks and actions to take.

6. Association with peers: It is an anthropological observation that you become like the people you spend a lot of time with.

7. Beliefs: What forms and sustains your attitude are your thoughts. Try to know others better. Every thought you have shapes your life. What you think about your neighbor is your attitude toward that person. The way you think about your project is your attitude toward that project. The sum of all your thoughts, filters, and assumptions comprises your overall beliefs.

8. Choices: Most people want to change the world to improve their lives, but the world they need to change first is the one inside them. That is a choice. You do not choose your project, your personality type, or your genetic makeup. Everything you are and nearly everything you do is not up to you. You must live with the conditions you find yourself in. However, you decide is how you do things. The longer you live, the more choices you make and the more responsible you are for how your life is turning out.

To change your life, to be more successful as a project manager, make a choice to take responsibility for your attitude and to do everything you can to make it work for you. Your attitude truly can become a difference maker. It is up to you. As in the following case studies, I needed to make a choice; it was not easy to do, but I did it.

CASE STUDY 1

I managed an IT infrastructure project in the North of Spain. The project involved two third parties and a team of twenty people. It was not a very complex project initially; however, it was a strategic project for the customer (Eusko Jaurlaritza) and also a very important project for the multinational company I represented. Furthermore, I was assigned as a project manager for this project just two months after joining the company. Everyone in that organization was looking to me. It was the first project I managed there and that represented a big challenge for me. I had six previous years of experience as a project manager in other companies. It was not very difficult for me to start the project and deal with people on the customer site. However, one month into the project assignment, my father was diagnosed with lung cancer. I was managing the project outside my city of residence, and I spent the whole week away from home. I felt very sad and stressed because of that situation, but I needed to make my choice. My first possibility was to resign from the project; but the consequences would be bad because everyone had high expectations about my job, the customer was happy with my work, and my team members felt loved by me. The second possibility was to continue managing the project but also to communicate the problem to everybody. To make my choice was not easy. My father was dying slowly. I remembered his words during my infancy: He said that ownership and commitment were so important for feeling good as a human being and also as a professional. He taught me always to separate people from the problems. Every time I was with him, he asked about my work, insisting to me: *Please do not worry about me.* Then I made my choice—I continued managing the project. Every weekend I was with my father at the hospital taking care of him. He never complained; he gave me love and positivism all the time. The project was moving forward. I had to manage third-party problems, technical issues, and people problems—nothing strange to me as a project manager. Six months later when my father died, my manager asked me why I did not tell him anything about my situation before. I never thought that I would be able to put up with my stress. I could not change my circumstances, but I could choose my attitude.

Not many people in my organization understood my attitude. I always say that we are not the center of the Universe. We live in an environment that continuously changes. We cannot change that environment, but we can change and adapt our attitude regarding our project environment. In that project, I could not allow people to worry about my personal problem. I needed to be the leader and to lead by example. I needed to smile frequently to my people who were away from their homes and families, too. They were looking and observing my behavior, and so I concluded that my choice was the only possible option. I do not like to be considered as a hero; it was my decision at that time. I refused any public reward from my company because I felt responsible for my decision.

CASE STUDY 2

During the past five years, I have been dealing with a personal project, getting my Ph.D. I had bad previous experiences at a couple of Spanish Universities, and I decided to be enrolled in a North European University, ISM at Lithuania. Most of my studies have been done virtually. I was committed to get my Ph.D. from the beginning of this project. In fact, after two failures, I found the right University to be enrolled in and felt happy. I followed a part-time Ph.D. program and received all the training needed to obtain the necessary knowledge and earn enough credits to qualify to move forward. I had a great supervisor who taught me how to write as an academic because I had some writing experience but as a project management practitioner. I got published my first research paper with my first supervisor, before starting my studies, and later got published two more research papers and presentations. Afterward, my university assigned me a new supervisor and I submitted two more research papers with her. But on January 24, 2023, I was diagnosed lung cancer, I could not have any surgery because I was in metastasis phase, so my oncologist told me the only thing they could do is to put an immunotherapy treatment aiming to reduce the tumors. It was like poring a cold-water jar on my neck. I could understand it, I was shocked.

I only had one year to finish my Ph.D. dissertation, and my mind told me, "Alfonso, move forward and get it done." My supervisor and my University Dean told me they would understand any decision I took; finally, I made the decision of moving forward. It has not been a path of roses, but my decision allowed me to stay mentally and spiritually strong. I am a catholic believer, and it is helping me a lot. Now on January 23, 2024, I defended my dissertation successfully and got my Ph.D. I had and still have daily secondary effects (insomnia, muscle pain, nausea, and tiredness), but I'm still alive, so I need to be grateful. Today is a great day!

Moving forward and getting my Ph.D. project finished allowed me to deal positively with my cancer. Now my cancer is stable, and I need to continue with my medication forever, but I need to be ready to deal with adversity with a big smile. Being focused on the positive every day helps me to move forward. I continue doing several activities; I am teaching some classes at the University, writing some articles, and mentoring some project managers. Those activities maintain my brain busy.

Many readers of this book have their own horror stories to tell. The lessons learned here are that we as project managers are responsible to choose our attitude. As the poet Maya Angelou says: *If you don't like something, change it. If you can't change it, change your attitude. Don't complain.*

SUMMARY

Attitude is an attribute that influences almost every action we might take as project manager. No matter how wealthy we are, how well trained, how successful, and whatever our station in life, we will finally be judged on the human element of attitude that we project. This chapter covers the concept of attitude. I say:

Attitudes are a secret power working twenty-four hours a day, for good or bad. Attitude is a brain filter through which you experience the world. Some people see the world through the mind of optimism, while others see life through a mind of pessimism.

The main ideas of this chapter are:

- Your attitude is your window to project success. We all start out in life with a good attitude.
- Wash your project window.
- Take charge of your attitude by keeping your project window clean.
- TODAY IS A GREAT DAY. Without a positive attitude, you cannot activate the other principles. However, the promise that attitude is everything is hollow. It serves as a difference maker.

This book addresses emotional issues, to help the project professionals cope with their leadership role, and to learn how to connect emotionally with the real payoffs that exist. To manage others successfully, you must manage yourself. Take the time to complete the self-assessment as follows. By doing so, you will end with an action plan that will help you face the challenges ahead with success and integrity. Make a plan to get what you want and start today.

TOOL: ATTITUDE ASSESSMENT

PM Attitude Self-Assessment Tool

Purpose

The purpose of this tool is to assess the attitude of the project manager. This tool provides a high-level analysis from potential attitude areas to improve. This is a first-level evaluation to sense the project manager attitude. Attitude is assessed regarding 8 critical success factors, based on the "Bucero's desired attitude model."

Remark

This tool may be used with project managers.

Administration

This survey can be used directly with the project manager or to be filled by an interviewer. The people who administer the tool need to know the attitude concepts.

Write a number from 1 to 5 as the most appropriate answer for all the following questions:

Scoring

1	Totally agree
2	Agree
3	Neutral
4	Disagree
5	Totally disagree

1. Respect

1.1	As a project manager, are you starting the day with a positive sentence in a daily basis?	1
1.2	Do you respect the people in your project?	5
1.3	Do you respect the team?	4
1.4	Do you respect the organization (procedures, management decisions, and guidelines)?	5
1.5	Are you reacting positively in front of bad news received from your team members or other project stakeholders?	3
	Total	**18**

2. Responsibility

2.1	Is your project manager clarifying the project mission and objectives periodically?	1

2.2	Is the PM asking questions to all project stakeholders in a disciplined manner?	5
2.3	Is your project manager dealing with the rest of project stakeholders in a correct and polite manner being receptive to their feedback and ideas?	4
2.4	Is the PM asking for feedback to all project stakeholders in a periodical basis?	3
2.5	Do you have a strong sense of responsibility?	5
2.6	Do you have a strong sense of commitment?	5
2.7	Have you spent time explaining to all project stakeholders your role and responsibilities?	3
	Total	**26**

3. Empathy

3.1	Are you a good listener?	1
3.2	Do you understand or make a special effort to understand others from where they come from?	1
3.3	Do you believe people like you?	3
3.4	Are you able to empathize with project stakeholders soon in the project life cycle?	3
	Total	**8**

4. Openness

4.1	When you need, are you asking your team members for alternative solutions?	2
4.2	Are you looking for applying previous lessons learned from previous projects?	2
4.3	Are you looking for ways to improve during a project?	2
4.4	Do you admit any team member opinion?	4
4.5	Are you open to admit any criticism from any stakeholder?	4
4.6	Are you open to talk to your Sponsor and Executives even when the project performance is not good?	5
	Total	**19**

5. 3P'S

5.1	Are you encouraging your project stakeholders and getting across positive messages?	2
5.2	Do you show up your enthusiasm to all project stakeholders?	4
5.3	Are you persistent when dealing with team members, executives, and other stakeholders?	4
5.4	Are you patient with all project stakeholders?	5
5.5	Do you show a positive smile every day during the project life cycle?	3
5.6	Can you put up with your stress being positive most of the time?	2
	Total	**20**

6. CII—Creativity, innovation, and intuition

6.1	Are you using your intuition in order to make a decision?	3
6.2	Are you creative when needed by the project?	5
6.3	Are you prepared to innovate when needed during the project life cycle?	5
6.4	Are you ready to follow the rules and guidelines but make an exception when it is justified in the benefit of project success?	4
6.5	Do you admit team member's creativity and innovation?	4
	Total	**21**

7. Willingness to participate

7.1	Are you willing to participate in your project?	1
7.2	Do you consider your contribution is important for project success?	2
7.3	Do you consider your contribution is necessary for organizational success?	3
7.4	Are you always ready to contribute?	4
7.5	Are you always ready to share your knowledge with the rest of project managers and executives in your organization?	5
	Total	**15**

8. Trust

8.1	Are you sincere, saying the truth about the project and dealing with your sponsor in a positive way?	1
8.2	Do you trust your organization?	3
8.3	Do you trust others?	5
8.4	Do you always report to your executives and other stakeholders the real situation of the project status?	5
8.5	Do you always say the truth to your team members and peers about the project?	5
8.6	Do you say the truth to your customer?	4
	Total	**23**

Scoring instructions

1. Add up the scores from each section

2. Divide the score by the identified factor to create an adjusted scoring

	Section	Scoring	Divisor	Adjusted Scoring
2.1	Respect	18	5	4
2.2	Responsibility	26	7	4
2.3	Empathy	8	4	2
2.4	Openness	19	6	3
2.5	3P's	20	6	3
2.6	CII	21	5	4
2.7	Willingness to participate	15	5	3
2.8	Trust	23	6	4

3. ADD UP the adjusted scores 27

4. Multiply total adjusted score by 2.5 to create the risk evaluation score 67

5. Select the correspondent cell to show
the risk evaluation

Global Attitude Evaluation		Precaution	High Risk
		41–70	71–100
Respect	18		
Responsibility	26		
Empath	8		
Openness	19		
3P's	20		
CII	21		
Willingness to participate	15		
Trust	23		

ATTITUDE EVALUATION TOOL

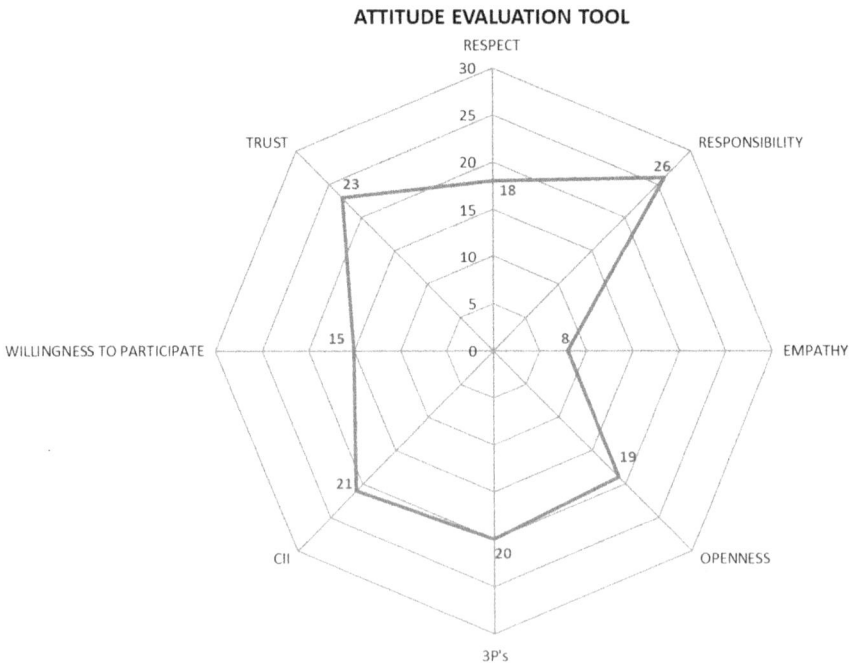

2

How to Attract Project Success

Success is not the result of spontaneous combustion. You must set yourself on fire.

–Reggie Lech

WE BECOME WHAT WE THINK ABOUT

I firmly believe that *we become what we think about*. I feel that our thoughts determine our actions. The project manager must be a believer from the beginning to the end of the project. If the project manager does not believe in the project, they will be unable to convince team members. Robert Collier offered this insight: *"There is nothing on earth you cannot have—once you have mentally accepted that you can have it."* A corollary thought especially relevant in the project management world is *you can have anything you want; you can't have everything*. In my talks and seminars, I use a popular Spanish joke expressing how important beliefs are for project managers: *A Spanish man meets a gypsy man, and the gypsy man says, "I have a horse to sell you."*

The Spanish man says, "A horse, for what? I do not need your horse." The gypsy man says, "Oh! It would help if you had it; it is a wonderful horse, wakes up early in the morning, and prepares breakfast for the whole family. It's a great horse."

The Spanish man says, "Ok, man, I don't believe you, but I'll buy the horse." Then, the Spanish man buys the horse.

Two months later, the two men meet again, and the Spanish man says to the gypsy man: "You were lying; it is an awful horse, it is crying all night, it is disturbing my neighbors, I don't want it."

DOI: 10.1201/9781003485674-2

The gypsy man answers, "You know, man, continue talking about the horse that way, and you will not be able to resell it."

We, as project managers, must sell our projects to organizations. First of all, we must start selling the horse ourselves. If we cannot believe in our horse, we cannot sell it. When you constantly think about a particular goal, you will take steps to move toward that goal. Let's say we have a project manager who thinks he can convince a customer about the benefit of a solution. Like a human magnet, that project manager will attract people to influence the customer to get confident about the benefit of that solution. The idea that we become what we think about is also known as the "Law of Dominant Thought." There is a power within each of us that propels us in the direction of our current dominant thoughts. The key word here is dominant. We have an internal power within each of us that moves us in the direction of our everyday thoughts. A little positive thinking does not produce positive results. It would help if you practiced thinking positively every day until it became a habit. You live your project by sending messages to your team members. You are preparing them for project failure or success. It is up to you!

ADJUST YOUR ATTITUDE

Let me share a personal example to demonstrate the power of our thoughts. When working as a project manager at HP Spain, I presented my first paper in English at an HP Project Management Conference in the US. I was not fluent in English at all that time; however, I was passionate about writing and submitting a paper in English. My dream was to be able to present my professional subject in English. I wrote a case study about my customer project in Madrid (Spain). The conference organizers answered me, asking for changes, and I did my best. I prepared my first English presentation for many weeks, and some weeks later, I went to San Jose, California, and presented my paper there to a worldwide HP audience. I was very nervous at the beginning of my presentation because of my level of English. However, I was getting better through my presentation. Why? Because I believed in that project and lived that experience with my team. At that time, my English level was not good enough. My passion for telling my story to the audience was the most important thing.

That presentation changed my behavior. Since that year, I have been sending papers to international congresses and symposiums, telling and sharing my experiences with different audiences every year, forcing myself to write and speak better and better in English. My good friend and professional, Randall L. Englund, is the mentor of my progress, who helped me progress professionally and personally and write better in English. Sometime after I submitted my first paper, he told me that acceptance was borderline; what swung the decision toward acceptance was the authentic message that came through—he could sense an attitude and story that needed to be shared. Thirty-one years later, I was a *PM Network* magazine Contributing Editor. I am the author of two Spanish and several English books, co-authored with Mr. Englund a book on *Project Sponsorship* (Jossey-Bass, 2006) and two more books (*The Complete Project Manager and The Complete Project Manager Toolkit*).

The books I wrote, co-authored with Randall L. Englund, were written with only three "face-to-face" meetings. Mr. Englund lives in Saint George, Utah (USA), and I live in Madrid (Spain). We exchanged many e-mails to write our book and took advantage of international project management conferences to review and agree on book material. Our project has finished, and we are pleased with our deliverables. One main lesson learned is enthusiasm for the project and believing in the possibility of making a change and contributing.

Another thing was doing it. My dreams have become a *reality* because I orient my thoughts to achieve a clear goal. This excellent experience has been fantastic for me. It was priceless because it taught me that you can reach your goal when you believe in yourself and keep your thoughts focused on the positive. You then attract others who want to work together with you. Furthermore, your organization will benefit from this situation, too.

ATTITUDE AND ACTION

I have been writing about the importance of positive thinking and positive thoughts. However, I have not written about where action fits into this process. We cannot achieve any results without actions. But thoughts and ideas precede actions; when our thoughts are positive, some actions immediately come to mind. My experience is that when I have a positive attitude, I feel compelled to take action, and nothing can stop me. It does not matter if I make some mistakes because I know I will be able to amend it. A positive belief system is the starting point for achieving any goal. When you think you can reach your goal, you begin taking the necessary actions to move forward. Let me give an example. When I wrote my first book, I worked as a Project Manager in a multinational organization. Sometimes, I had long journeys, and I started to write after dinner, which I considered a daily task. Some days, I was too tired and could not write a word. Other days, I spent two or three hours writing down my thoughts. I always needed to see something tangible and accomplished little by little. I printed out my work, encouraging me to continue until I finished my book. I believe managing projects in organizations works with a similar approach. The project manager and their team need to achieve small deliverables as soon as possible because it generates motivation among the team.

YOUR CIRCUMSTANCES

We are sometimes not conscious that we create a lot of negativisms every day in our environment. Your beliefs brought you to where you are in your career today, and you're thinking from this point forward will take you to where you will be in the future. I have wanted to be in the project management for many years. I started my career as a programmer, software engineer, and team leader until I achieved my first goal (to be a project manager). I was lucky because the organizations I worked for allowed me professional development. The activities I participated in over many years were because I dreamed of being a project manager and ultimately achieved that goal. My desire to be a project manager moved me to ask project managers about what they did and to learn more about the profession. That helped me to develop my professional development plan and my vision. It was the way I found Project Management Institute and discovered the vast number of opportunities the project management profession has. It is keeping me in touch with more than 700 worldwide project professionals over the last thirty years.

I was lucky to meet Randall L. Englund in 1992 in Santa Clara (California), and he always encouraged me to put my dreams into action. We both worked for Hewlett-Packard, and he supported me from the Project Management Initiative at Hewlett-Packard in Palo Alto over the years. As professional chemistry worked among us, we started submitting articles to several PM Conferences, and over the years, we published four books together. We exchanged our positive experiences and learned from them. He has been my mentor and belongs to the Positive Professionals Club. I know some people for whom these trends are innate, and others are not. However, changing your attitude about your career and life is possible. I believe project managers must be ready to serve their team members, customers, and a variety of stakeholders.

CHANGE YOUR THINKING

What are you thinking about yourself every day? Do you have positive or negative thoughts? Each of us has an internal voice. Many times, what we say is negative, critical, and self-limiting. We create our obstacles; nobody else does it. Perhaps you think, "I cannot do this," or "I always mess things up." These thoughts work against you. Instead, repeat to yourself that you can and will accomplish your goal, project, or activity. Many times, we are not taking into account the words we use regularly. We continually use negative ideas to express ourselves—Don't turn right. Do not think negatively. These statements make our point consciously, but how are these innocent messages being received at the unconscious level?

Every time you use a negative, your brain first registers the exact opposite of what you mean. For example, we met a friend with a broken toe the other day and asked her what had happened. She replied, "I was shopping and had just picked up a tin of paint when a salesperson said, 'Careful, don't drop it on your foot.' A few moments later, the paint slipped out of my hands." The unconscious processed the helpful salesperson's message as "Drop it on your foot—Don't," and she followed the initial command to the letter!

How often are children disciplined with negatives? Don't fall. Don't touch that. Don't, don't, don't. Sometimes, what may seem like disobeying could be their unconscious processing of the experience of falling or handling. Have you ever thought, "I don't want to fail," "I don't want to be in this job all my life," or "I must not miss this sale?" Think about it: By the time your brain has got to the "don't" or "must not," your whole body has already received the message: fail, stay in the job, and miss the sale. So, if you want to think positively, the critical question is: "If I don't want" X, "what do I want in its place?"

Answering this question gives your brain something to work toward. The more you specify what you will see, hear, and feel, the more the brain is given a map to create new solutions and guide it to its outcome. So, if you think something like, "I don't want to be nervous in front of a group," ask yourself, "What would I be doing instead?" You might answer, "I will have steady hands, breathe easily, feel relaxed in my stomach, make eye contact, have a clear voice, be humorous, etc." The more detail and the more specific you are, the better. Your brain needs a particular goal and a DIRECTION to achieve that goal to be effective. If it does not have a direction, it wanders randomly, avoiding what you don't want but not going anywhere. You can aim in a specified direction and maintain that course, or you can say, "Good luck"—and indeed, whatever will be will be—but it will be outside your control. One way to get used to moving in a desired direction is to continually ask yourself, "What do I have to do NEXT to achieve my goal?" As far as the unconscious mind is concerned, the goal is like a map. The direction is like a compass. You need both to navigate successfully. Strategies for motivation use the mind's ability to move you toward what you want and away from what you don't like. I recommend using positive words and comments about yourself as a project manager and your goals. In Figure 2.1 are suggestions to help you become more positive and get the results you want:

- Every day, I read some positive literature. Find fifteen to thirty minutes in the morning to do this task. If you have the opportunity, do the same before going to bed.
- Every day, I listen to motivational cassette tapes. The key is repetition. When you repeatedly hear these messages, they become part of you—and you begin implementing them to improve your life.
- Join positive people. Look for positive and enthusiastic professionals and establish a professional relationship with them.

Do it every day
Find 15-30 minutes to do it every day
If you have the opportunity do the same before going to bed

1. Read some positive
LITERATURE

The key is repetition
When you hear those messages over and over they become part of you

2. Listen to motivational cassette tapes

Suggestions to help you to be more POSITIVE

3. Join positive people

Look for enthusiasthic professionals
Create a relationship with them
Keep your network up to date

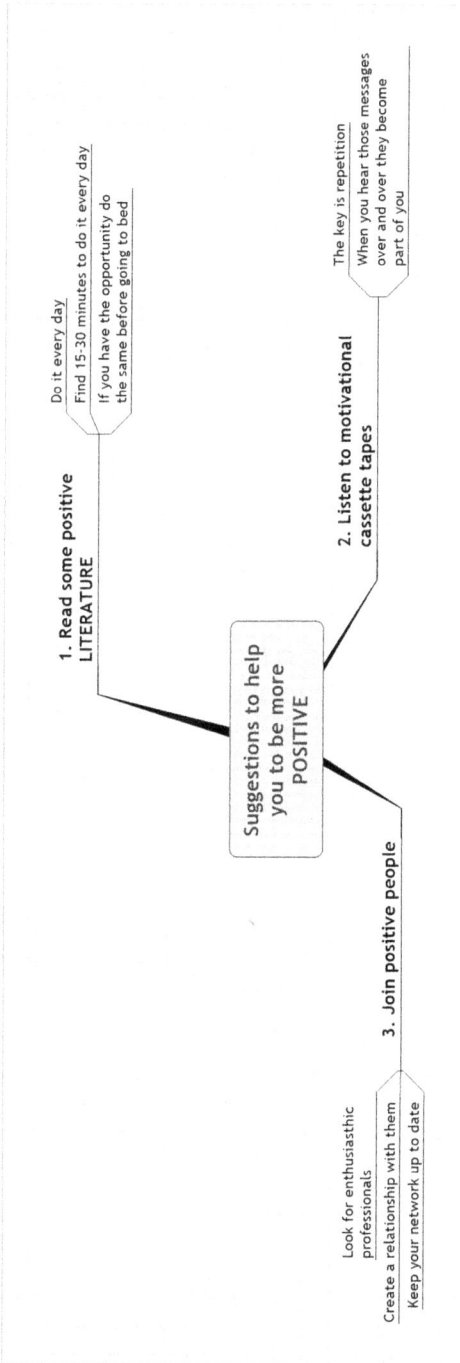

FIGURE 2.1
Suggestions to help you to be more positive.

If you follow this advice, it will make a phenomenal difference in your life. I can tell you that these techniques work if you have the discipline to stick with them. My lesson learned is: Change your thinking, and you will change your results in your activities, your projects, and your life.

OVERNIGHT PROJECT SUCCESS

Positive thinking is a process that takes time and patience; it is not an overnight success. Positive project thinking requires effort, commitment, and patience. On the other hand, positive thinking does not mean the absence of problems. You will have a lot of setbacks along the way. However, if you continue to believe in yourself and be persistent, you can overcome those obstacles. Believe me; you are constantly moving toward your dominant thoughts. Everything you achieve in your career flows from your thoughts and beliefs. Negative thinking yields negative results, and positive thinking produces positive results. It simply makes no sense to think negative thoughts unless you want to get negative results. Then, from this point forward, choose your thoughts wisely and use these experiences to get fantastic results in your career and projects. These ideas regarding optimism may help you throughout a project life cycle:

Calm down—Don't expect everything to work your way. Life is a wheel; chill out whenever you are on the wrong side of the turn. Go out and find outlets to release pressure and negative feelings. Forcing the issues won't likely work.

Downplay, but do not take for granted—If things go wrong, downplay them to ease the negative emotions. If you cannot join an outing or important gathering because of prior non-enjoyable work, assure yourself of the benefits you will get from work, and the meeting won't be as fun as you thought it would be.

It's not the people—Processes and events should be top priorities. Do not blame yourself when something wrong happens. Look at events and other external factors that could have caused the incident. If you cannot accomplish something, do not tell yourself you are not good enough; instead, recognize you need more training and education to deliver at high levels. Give yourself the benefit of the doubt and be thankful that you are a living being getting exposed to a variety of experiences.

Stand-up—Don't be afraid to fail and stand back up. Optimistic people learn and use their failures to get back on track. Reflect on what's happening and how to learn from it.

Same feathers flock together—Make sure you hang out with optimistic people. Their attitude will help you stay positive and encouraging.

CASE STUDY 1: CREATING A PMO

In the Solution Business, communication and documentation with the client and within the delivery organization are critical to the project delivery process. The Project Manager needs to manage customer expectations to get things done. They must be relieved of many internal organizational concerns to maintain this focus. The difficulty increases when a project culture does not exist—or what is there is very weak—in an organization that delivers customer projects. The management team only focuses on numbers and results. They ask for good project results but do not worry about how to establish and create the right environment for project success. In this kind of organization, upper managers do not support the project manager, and project managers are on their own when dealing with internal and external stakeholders during the project life cycle. Sponsorship was an unknown term in this environment. A project office implies some innovation because it changes the way to proceed—creating the ability for the project manager to stay focused on the client and perform high-quality project management. The project manager must analyze all internal and external stakeholders and their expectations and assign a very effective and empowered team.

Background

The foundation for the HP Spanish Project Office began in September FY99. As the organization grew regarding projects and people, knowing more about project status became a real issue from management's perspective. Then, the management team decided to implement a Project Office to relieve project managers of low-value activities. They asked me to work on a "solution proposal." HP presented this proposal, which the management team studied, discussed, and finally accepted in February 2000. The Project Management Office (PMO) project started at the Madrid office on

March 1, 2000. The Spanish Project Office arose from the need to relieve project managers of administrative tasks associated with managing a project. The Hewlett-Packard Consulting Organization believed the Project Office should help the project manager improve efficiency, facilitate getting the right tools, and align services with the needs of the project environment. Generally, at a corporate level, project offices are considered project management centers of expertise. HP decided that the professionals who staff these project offices should be experienced and trained in project management skills. At the local level, we use the Project Office to change our culture from a reactive style to a project-oriented organization. This culture shift predominantly occurs by demonstrating or modeling the new behavior. Our approach is to get initial results on a subset of projects and use these successes as models for others to adopt our process. It was a step-by-step process of achieving small, tangible things in the organization.

Mission and Objectives

It took some weeks to get an agreement with the management team about the reason for this particular project. Why do we need a Project Office?

I explained to the management team that Project Office adds value to project team members by providing mentors, consultants, training, tools to be more effective, and structured intellectual capital. I was always enthusiastic and positive with my comments, talks, and discussions. The Project Office adds value to the organization, providing a culture shift to project management, reusable tools and techniques, document and methodology support, global recognition, profitability improvement, and quality support. The Project Office also adds value to our customers, providing visible signs of HP commitment, competent HP team support, and more effective and quick answers.

Critical Success Factors

In all the projects I managed in my professional career, I found that project success depends on *how well you work with and lead people.* That behavior may attract people to achieve project success. I learned that aligning the project office and the organization's culture is necessary. Engineers may solve technical problems, but it is different when discussing people interactions and team members' relationships. Although we identified some factors as critical in the PMO implementation project, one of the most important is to focus on being prepared to answer questions and demands from different project stakeholders. Each consultant and PM expect the PMO to help them daily, which means being ready for uncertainty. Frequently, the type of demand generates pressure in terms of time or expectations, and we, as the PMO, need to transmit feasibility and security. *I always asked for proactive behavior from each PMO team member. I discovered* the following success factors:

- Scope agreement and clear setting expectations between all users and stakeholders (it took some weeks of meetings and validations)
- Forming, storming, and norming the PMO team (subcontracted people developed 80% of my team. I employed time for initial training, methods, tools, and procedures)
- Clearly defined functions, roles, and responsibilities for the PMO (I talked one-on-one and verified each person's expectations)
- Sponsorship from the upper-level management (I asked the General Manager to invite people to use the project office services)
- Clear communication plan deployment
- Communicate project status periodically to the management team and the end users
- Positive attitude, team service orientation

Lessons Learned

I learned a lot from the PMO project. One of my essential behaviors was transmitting passion and positive behaviors to my team members and executives.

- A PMO team needs time for:
  - Forming: It takes a lot of effort and time. I need to invoke patience many times.
  - Storming: Allowing people to generate and brainstorm ideas and thoughts.
  - Norming: Discipline and guidelines are necessary for all types of people but are mandatory for inexperienced team members.
  - Performing: I always gave my team the benefit of the doubt.
- People will support a project office (and communications in general) when they see its value and how it links directly to positive business impact.

- Finding and provoking energy in your team members is significant to achieve project success. Empowering my team members was necessary to manage team and project issues.

CASE STUDY 2: GAINING EXECUTIVE SUPPORT

This case study describes a Spanish Savings Bank that decided to implement a project portfolio process in its IT organization. The PMO manager from the Savings bank led the project with the help of a PM consulting company. They worked on a team with the savings bank professionals to sell the Executives the need for a project portfolio process and the added value of this process implementation for their organization and their customers. This case study deals with the problems, difficulties, and lessons learned from this challenging project.

Background

As the organization had been growing in terms of projects and people, knowing more about project status became a real issue from the management perspective. Then, the management team implemented an IT Project Office to relieve the project managers of low-value activities. After two years, they selected my organization (BUCERO PM Consulting) to help a savings bank located in northern Spain implement a project portfolio management system in their organization. The customer was called "Caixa Galicia" in the North West of Spain. They had a PMO in place but only focused on control. The customer used the PMO to prepare progress status for the Management of the Bank, but they did not gather enough information about the projects they manage in their organization. I worked as a PPM consultant, and my first activity was a project portfolio assessment. As a result of that assessment, I discovered that they were doing too many projects, so there were:

- No holistic view of project portfolio in the organization
- A mix of senior project managers and outsourced people
- Lack of discipline
- Lack of sponsorship
- Focus on control

Upper management felt they invested too much in projects that did not add value. They invested effort, time, and money in projects with very little added value, which they needed to do to operate more efficiently as a banking organization. The urgency of implementing a project portfolio management system was clear enough.

Executives' Beliefs and Expectations

Caixa Galicia Executives did not have a high level of project management maturity. They believed that the PMO was helpful for project control, that it was tactical, focused on launching new products on time, and was first to market. I looked for the opportunity to talk face-to-face with the General Manager of that organization, and after several tries, I achieved it. The General Manager did not see the project management added value; he thought about project management as a tool.

He could not create the right environment for successful projects in his organization because he did not see that need. It was curious because that professional had been working for a consulting company in the past before managing the IT department in the Bank. However, his way of managing people in his organization created a lot of interruptions, so project managers and technical professionals were primarily involved in firefighting. Project priority changes were so frequent. As a consequence, middle managers followed the same behavior. The General Manager focused on control instead of results. He asked people to set up priorities but changed them frequently. I spent some time explaining the project portfolio management basics to him and how that approach would help his organization to be more efficient and better organized. The General Manager asked me to educate middle and business managers on project portfolio management because they lacked knowledge. Then, I prepared a training workshop for middle managers and executives.

When I started the session, they asked me why they needed a portfolio management process. They said that they had been successful so far, so why change? I answered them that the process would help them decide which projects to invest in, allow them to monitor them, and make decisions about moving forward or canceling projects. They also asked me about the cost of implementing and running a project portfolio process, and I told them that it would always be less money and effort than not having one. Finally, we discussed the services and added value for that system. I remember it as a very challenging meeting for me as a consultant. However, it benefited everybody because the workshop helped them identify the current situation, and they came up with the idea that they needed to change. I got excellent feedback from the seminar; I got comments like: "It has been great to find somebody who spent time with us talking about strategy and portfolio management without any interruption."

The Key to Getting Upper Management Support

The key to getting upper management support was to show how a project portfolio system would help them solve current problems and provide business impact. They did not have a standard list of projects and programs in the organization. Management knew about some projects but did not know if they were investing in the right ones. They invested a lot of money but did not see the risk. I asked the General Manager to be the project sponsor. The success or failure of any project often hinges on how well the project sponsor—the person who funds the project and ensures that it achieves the desired benefits—relates to the project, the project manager, and other stakeholders. However, executives assigned as project sponsors often have little, if any, experience understanding their roles and responsibilities during the project life cycle. Problems in communication and execution are inevitable as long as senior managers and project managers do not comprehend the mechanics of their relationship. And I ran a Stakeholder by asking the project team the questions as follows:

- Who are the stakeholders?
- What are the stakeholder expectations?
- How does the project or product affect stakeholders?
- What information do stakeholders need?

Once I got their answers, I classified them and organized my project stakeholders in a matrix using the below-mentioned tool (Figure 2.2). The bubbles show the stakeholders; each bubble's size represents the amount of project support. You can see in the diagram that around 45% of the stakeholders had high power and interest in this project.

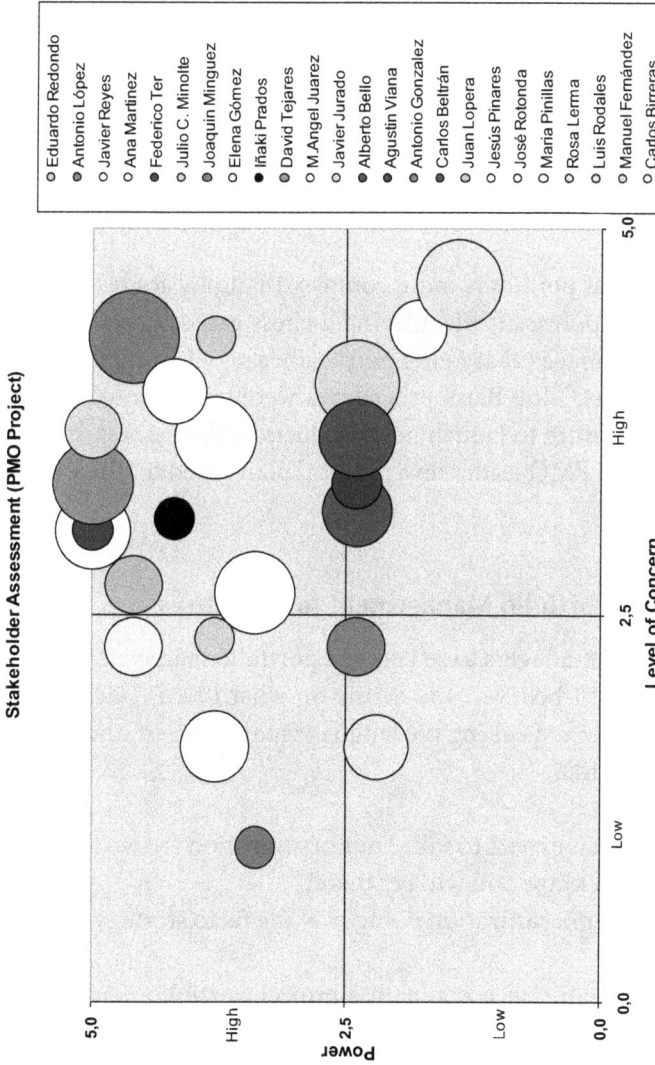

FIGURE 2.2
Stakeholder assessment.

The Project Portfolio Project Stakeholders

I was lucky because the PMO leader could influence people and help me convince them about the importance of that project and its value to the organization. In any case, the resistance level was initially because every stakeholder believed they were critical; very soon, requirements prioritization became a big issue. I needed to be a "good preacher and bullfighter." I needed to clarify roles and responsibilities for all the team members, but I needed to be flexible with the whole organization. Managing politics during the Project Portfolio project on the customer site was challenging. Some managers only managed "win–lose" situations. All business units believed they had the same strategic weight in the organization.

I always say that politics is more complex than physics because we can formulate it, but politics is like playing a chess game. I coached the customer project manager that being politically savvy in organizations and projects is a "must." The Bank procedures were tricky to follow, and the management pressure to launch new products to the market was so high. I worked with the PMO leader on a political plan based on the stakeholder analysis we did.

Selling Project Portfolio Management to Executives

I spent time with managers to sell project portfolio management to executives, saying what I believed and acting on what I said. I delivered some talks and workshops speaking the language management understand and asking questions like:

- Where do you expect to take your organization in two years?
- How do you know you will get there?
- What is your organization doing now (de facto strategy)?

I explained to them that a company's project portfolio drives its future value. Successful strategic execution requires tightly aligning the project portfolio to the corporate strategy. I explained to them the key was translating the strategy into the project portfolio. I proposed to the customer a process to sell sponsorship into the organization. You can see the set of steps I followed below. Individuals fulfilling the sponsor role must reinforce sponsorship excellence's features, advantages, and benefits.

Assess

Understand the need to have a sponsor assigned to your project. Understand the need: Focusing on the organizational need for a project sponsor is the first step in the process. Each level in the organization will have a list of its priority needs. The best way to cover the executive needs is to understand what the key strategic priorities are for the organization. Executives could obtain strategic priorities through periodic reports, discussions with senior executives, and meetings. Understand their strategic priorities, goals, and objectives, and see where your project is and how important it is for your organization.

I strongly recommend assessing your environment (Project Environment Assessment Tool). Then, you must select a sponsor that will be most affected by the benefits and value produced by the project. Find a sponsor that is responsible for a critical business. Suppose this same individual was also accountable for troubled projects from the past. In that case, gaining their support will be easier when you can relate to those projects and explain how a strong relationship between the sponsor and the project manager can produce a significant business impact. Share your findings with other managers in the organization, and always speak about the project and its business impact, not about you as a project manager believer. Ask the management team for consensus about assigning a sponsor to your project.

Plan

Involve your assigned sponsor from the beginning, and talk about their and your expectations. Remark on how the project sponsor can help the project manager and their team achieve a significant business impact throughout that project. Brainstorm what, how, and when the sponsor will be necessary and their interaction with the project manager. Develop a communication plan. People often forget to communicate with business leaders and those needing project management. The plan must include the target audience, frequency, and type of information presented (issues to mitigate, escalation process, progress updates, capabilities, and benefits). Advertise, advertise, and communicate; you must sell sponsorship. Prepare a Business Case: All key business unit personnel must complete a business case. Producing the business case as a team will help get buy-in from all departments and upper managers. This business case will be your selling tool for gaining funding approval.

The business case must include:

- Key business challenges, goals, and objectives you want to address
- Proposed sponsorship strategy. Your approach, your expectations, project resources
- Benefits and value project sponsorship will bring to the organization;
- Proposed cost
- Roll out plan

The secret to selling an executive is to focus on the organization's primary business needs and the value sponsorship can bring. The primary business needs come from the first step (understanding the need). The two essential value items that executives want you to focus on are 1) how project management can reduce costs and 2) how project management will increase revenues.

Follow

Talk to your sponsor frequently. Keep them informed about the project. Help them understand that by working together, the sponsor will know much more about the project and the customer, and they will be much more valuable to the organization. Use management meetings to reinforce and underline the support and help of your sponsor. Explain the value to the rest of the managers.

THE EXECUTION

The project portfolio—the array of investments in projects and programs this company chose to pursue was the agent of change. I provided them with small but tangible results as soon as possible. We did several actions:

Executive involvement in project reviews: The PMO randomly called for a meeting with some project managers and executives. They sat down together to run a project review. That meeting was an excellent opportunity for the executives to learn more details about the project and to understand the project manager's difficulties and lack of skills to improve project management in the organization.

Project portfolio training for all the project stakeholders: In that training, the PMO team helped me gather information about the projects in the IT Bank portfolio. In that way, I could understand the level of knowledge executives had about the projects they sponsored in the organization.

Half-day training for business managers: In that session, I explained the concept of project sponsorship, its meaning, and its implications to all business managers. I spoke the language management understands.

General Manager coaching in project sponsorship: It took some time and effort, but the General Manager appreciated the value added at the end of the project. He understood the role of the sponsor very well, and it was one of the reasons for the project's success.

From the beginning of the project, we obtained small results to show to management. It was one of the critical elements to be successful. The PMO played a vital role in the project, managing project stakeholders. The PMO leader knew all the main stakeholders well, which helped me immensely when I proposed taking some actions. The behavioral change from the sponsor step by step during the project life cycle positively influenced all project stakeholders.

THE PROJECT PORTFOLIO MANAGEMENT TRAINING

The training focused on the key following subjects I considered:

- All projects need a sponsor
- Project Sponsor role and key responsibilities
- Understanding organizational culture (strategic alignment and business impact)
- Commitment and ownership
- Setting and maintaining the agreed-upon priorities

We reviewed all the types of projects and demonstrated that every project would need a sponsor to succeed. We defined the role and responsibilities of the project sponsor in Caixa Galicia. We created a cultural assessment to determine the strategic alignment and the business impact.

We explained the importance of commitment and ownership from the executive point of view, and we remarked on the importance of setting up and maintaining project priorities.

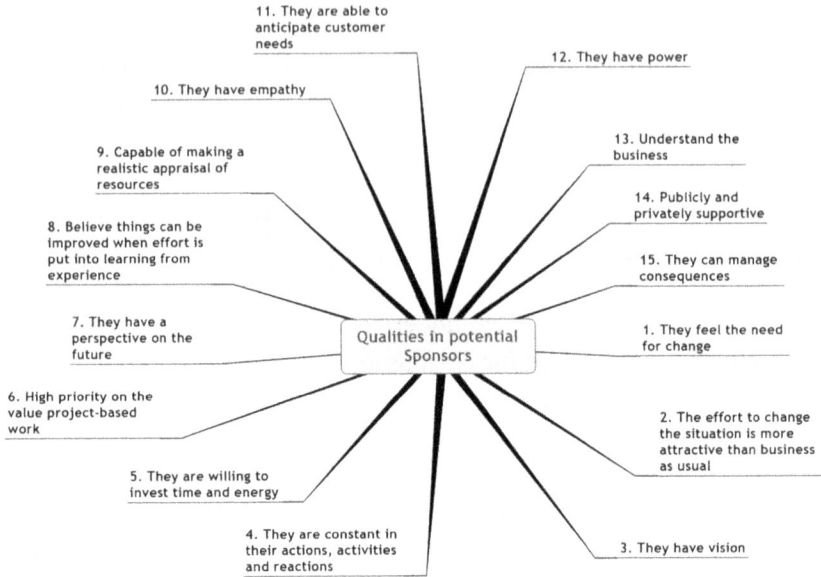

11. They are able to anticipate customer needs

12. They have power

10. They have empathy

13. Understand the business

9. Capable of making a realistic appraisal of resources

14. Publicly and privately supportive

8. Believe things can be improved when effort is put into learning from experience

15. They can manage consequences

7. They have a perspective on the future

Qualities in potential Sponsors

1. They feel the need for change

6. High priority on the value project-based work

2. The effort to change the situation is more attractive than business as usual

5. They are willing to invest time and energy

4. They are constant in their actions, activities and reactions

3. They have vision

FIGURE 2.3
Qualities in potential sponsors.

As shown in Figure 2.3, we explained some criteria for selecting the right Project Portfolio Sponsor.

After the training, the Executives understood Sponsorship implementation may have a significant business impact. Effective sponsorship can contribute to increased business success. Sponsors implemented the plan agreed upon with the General Manager. Finally, the PMO followed up with the General Manager. Caixa Galicia was a functional organization with a very autocratic management style. It was not easy to change those behaviors step by step. I had to use my passion, persistence, and patience during the project's life cycle. Although I had a lot of enthusiasm acting as a consultant, the project manager was the PMO leader, who had worked more than twenty-five years for Caixa Galicia and knew very well the procedures, the politics, and, more importantly, the people at all hierarchical levels in the organization. He acted as a change agent in this challenging project.

HOW TO SUSTAIN THE PROJECT PORTFOLIO DISCIPLINE

A significant problem with project sponsors is to keep them involved and committed to the project during the complete project life cycle. From personal experience in software and infrastructure customer projects, I find that the sponsor of the delivery organization was very committed from the beginning of the project until the project sale. Then, the sponsor disappears. One reason for this behavior is that project sponsors do not realize the need for continuous project sponsorship from beginning to end. Other reasons include different personality styles, lack of knowledge, priorities, and lack of interest. The sponsor may not have been very committed to the project in the first place or may have been too busy doing too many other things. When measuring sponsors by sales objectives, then when the sales are complete, they move on. Perhaps it did not occur to them that by continuing as project sponsors and staying active, they would know the customer much better and be able to sell more. The measurement system may focus only on short-term, silo-based activities and reward efficient individual efforts instead of optimizing organizational throughput and accomplishment. The sponsor can bring perspective, reinforce the big picture, and fortify or diminish project efforts based on strategic goals (Figure 2.4).

Challenges to address: As a project sponsor, the activities to perform vary during a project life cycle. The level of involvement is also different. Project sponsors better serve project managers and their teams when they

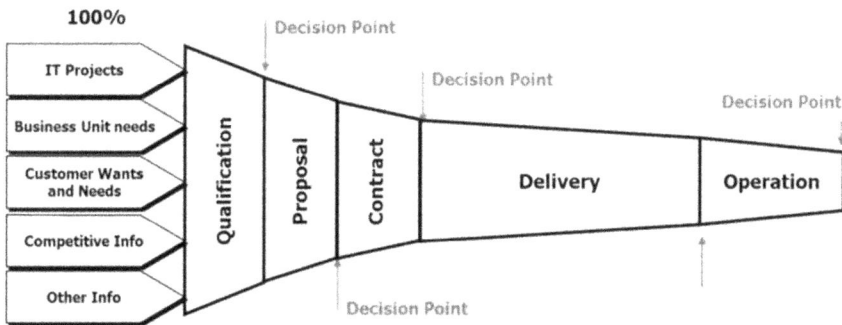

FIGURE 2.4
Project portfolio selection.

support and do not interfere. During the initiation and planning phases, the sponsor plays an active role in helping to establish project objectives. The sponsor guides the project manager in making decisions during organization and staffing phases. The project sponsor is probably more familiar with organizational politics and can help to navigate the political factors that influence project execution.

A problem that often comes up is changes to project priorities. The project sponsor can work alone or with other executives to agree on project priority and then inform the project manager, explaining why they assigned that priority. Project sponsors are managers or upper managers who know or should know how the organization works. Then, the project sponsor can help the project manager establish processes and procedures for the project—the project sponsor functions as the contact point for customers and clients. Project portfolio reviews helped the organization force executives to meet project managers periodically to review the project status and see the problems and issues their project managers faced. Then, we moved forward through project sponsorship. These activities reinforced the project sponsorship culture in the organization.

Proactive sponsorship: The ideal situation is proactive sponsorship—getting a committed project sponsor who is accountable, takes the project seriously, is knowledgeable, is trained, and is able not only to talk but also to walk the walk. The person is trustworthy in all respects. Their values are transparent and aligned with the organization and its strategy. This sponsor protects the team from disruptive outside influences and backs them up when times are tough. Rather than correcting a deficient sponsorship situation, a preferable way is to start right in the first place.

Project reviews: To create a good relationship between the project manager and the project sponsor takes time and needs fine-tuning. One method we found useful and strongly recommend is to run monthly project reviews led by the project sponsor. Those meetings add value to the project manager, sponsor, and the organization. They force the project manager to review their project status and pending tasks. At the same time, it drives the project sponsor to know more about the project, the customer, and other stakeholders. In the solution business, for example, that practice helps the project sponsor know the customer much better, generating more business. To do those project reviews, I prepared an Excel spreadsheet filled in for every project reviewed. The questions asked in those reviews were in Figure 2.5 as follows:

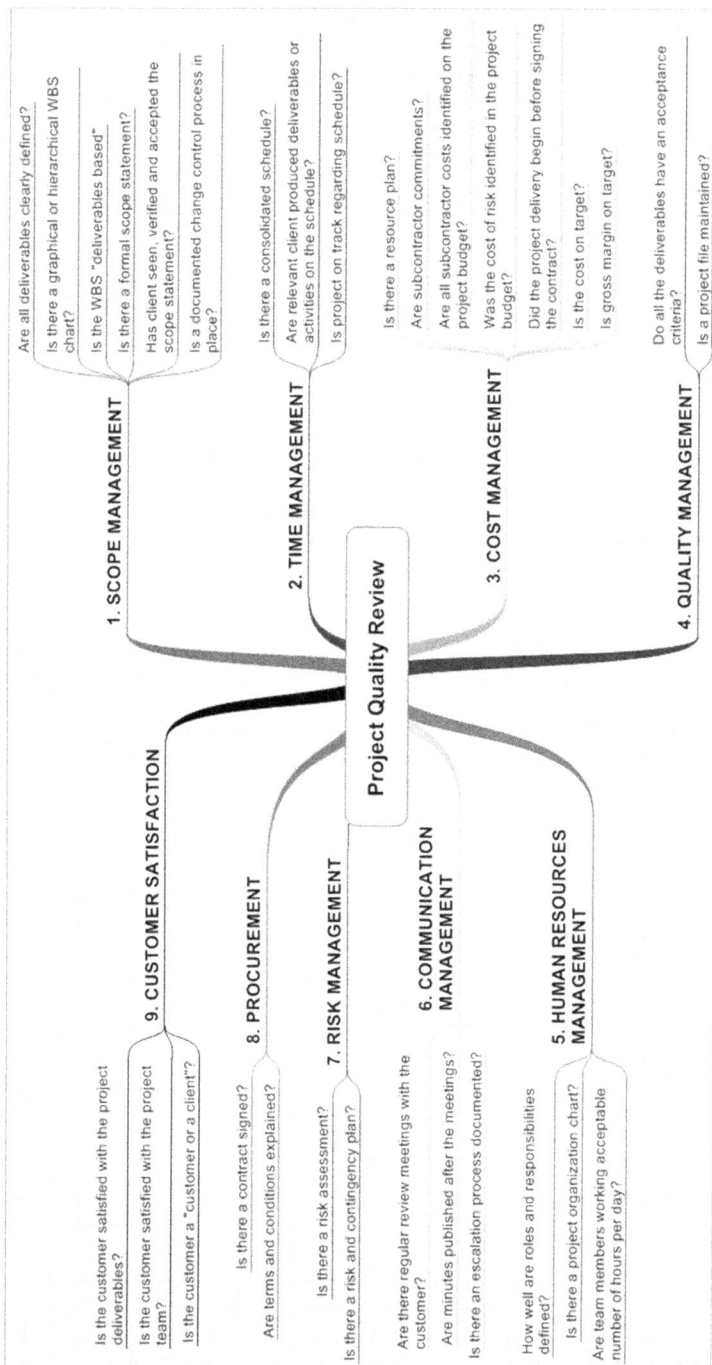

Are all deliverables clearly defined?

Is there a graphical or hierarchical WBS chart?

Is the WBS "deliverables based"

Is there a formal scope statement?

Has client seen, verified and accepted the scope statement?

Is a documented change control process in place?

1. SCOPE MANAGEMENT

Is there a consolidated schedule?

Are relevant client produced deliverables or activities on the schedule?

Is project on track regarding schedule?

2. TIME MANAGEMENT

Is there a resource plan?

Are subcontractor commitments?

Are all subcontractor costs identified on the project budget?

Was the cost of risk identified in the project budget?

Did the project delivery begin before signing the contract?

Is the cost on target?

Is gross margin on target?

3. COST MANAGEMENT

Do all the deliverables have an acceptance criteria?

Is a project file maintained?

4. QUALITY MANAGEMENT

Project Quality Review

Is the customer satisfied with the project deliverables?

Is the customer satisfied with the project team?

Is the customer a "customer or a client"?

9. CUSTOMER SATISFACTION

Is there a contract signed?

Are terms and conditions explained?

8. PROCUREMENT

Is there a risk assessment?

Is there a risk and contingency plan?

7. RISK MANAGEMENT

Are there regular review meetings with the customer?

Are minutes published after the meetings?

Is there an escalation process documented?

6. COMMUNICATION MANAGEMENT

How well are roles and responsibilities defined?

Is there a project organization chart?

Are team members working acceptable number of hours per day?

5. HUMAN RESOURCES MANAGEMENT

FIGURE 2.5
Project quality review.

CONCLUSION

There are some thoughts that I would like to share with you based on my experience in this project:

- Selling project portfolio management to executives and getting buy-in needs time and effort.
- Politics is active in every organization.
- After training executives in Project Portfolio and Sponsorship, you must follow up on the manager's action plan to achieve good results.
- Selling project portfolio management to Executives is dealing with power in organizations.

SUMMARY

I believe that our thoughts determine our actions. The project manager must be a believer from the beginning to the end of the project. As project managers, we must sell our projects in organizations, especially if you are a project manager who thinks you can convince a customer about the benefit of a solution. Like a human magnet, you will attract people to influence the customer to get confident about the benefit of that solution.

Please remember these critical ideas about positive thinking:

- My experience is that when I have a positive attitude, I feel compelled to take action, and nothing can stop me. It does not matter if I make some mistakes; I do because I know I can find and implement a solution.
- We are sometimes unaware that we create a lot of negativity daily in our environment with our comments and reactions, and people always observe us as leaders.
- I recommend using positive words and comments about yourself as a project manager and your goals.
- Positive thinking is a process that takes time and patience; it is not an overnight success. Positive project thinking requires effort, commitment, and patience.

So, how can you attract Project Success? Before moving to Chapter 3, please do your self-assessment. Start today.

TOOL: ATTRACT PROJECT SUCCESS ASSESSMENT

As a project manager, you can assess how much you attract when managing projects. Please reflect upon the following questions:

1. Feelings

   - Do you feel compelled to take action, and nothing can stop me? (It does not matter if I make mistakes; I do because I know I can find and implement a solution.)
   - What do you think about you as a project manager (do you feel ready)?
   - What do you think about your project?

2. Circumstances—Word Awareness

   - Are you aware of the negative words you are pronouncing every day?
   - Are you making positive comments about yourself and the project you are managing?
   - Have you analyzed your circumstances?

3. Take Action

   - Are you applying your patience when managing projects in organizations?
   - Are you able to adjust your attitude?
   - What are you doing to improve?

Based on those questions, you can prepare your action plan. Remember that you are an excellent project manager, then you can do it.

3

Make Your Plan for Success

You must first clearly see a thing in your mind before you can do it.

–Alex Morrison

Since the age of twenty six, I dreamed of being a project manager. I started to work in the Information Technology field as a programmer, then as an analyst, and then as a project leader, leaving and joining several organizations. As many readers can attest, I worked hard to earn every bit of my position. Now, I have my firm and work on my passion, project management. I believe you must dream before achieving things because when you desire, you unconsciously generate synergy for the steps to follow to achieve your objectives. Word class swimmers also incorporate the power of imagery to reinforce in their minds how they want to perform. Many top competitors mentally envision a successful outcome before achieving it in the "real" world. Visualization refers to seeking to affect the outer world by changing one's thoughts. Creative Visualization is the basic technique underlying positive thinking. The concept arose in the US with the nineteenth-century New Thought movement. One of the first to practice the method of Creative Visualization was Wallace Wattles (1860–1911) (2022), who wrote *"The Science of Getting Rich,"* which he believed was based on spiritual principles. Creative Visualization is using your imagination to create what you want. You need to do your Creative Visualization in the first person and the present tense—as if the visualized scene were unfolding all around you, whereas normal daydreaming happens in the third person and the future tense—the "you" of the daydream is a puppet which the real "you" is watching from afar. Visualization

DOI: 10.1201/9781003485674-3

practices are a joint spiritual exercise, especially in esoteric traditions. In Vajrayana Buddhism, for example, complex visualizations are used to attain Buddhahood. Visualization is not reserved solely for actors/ actresses, athletes, or movie stars. You have used it since childhood to create the circumstances of your own life. I describe Visualization as "movies and pictures of the mind," "inner pictures," or "images." We all store pictures in our minds about the type of relationships we deserve, the degree of success we will attain at work, the extent of our leadership ability, and so on.

I observed that we develop "inner pictures" early in life. If we were criticized or felt unworthy, as youngsters, we recorded the events (and the feelings associated with those events) as images in our minds. Because we frequently dwell on these pictures (both consciously and subconsciously), we tend to create life situations that correspond to the original image. For example, you may still hold a vibrant image of being criticized by a manager in a project meeting. I still remember one manager I worked for as a project manager in a multinational company; he always gave me negative feedback. He never gave me a positive word. He reprimanded me in front of my team members. He made me feel terrible and decreased my self-confidence. The picture remains in your mind even when you are not always conscious. These pictures exert tremendous influence over present actions or project activities.

TAKE OWNERSHIP AND CREATE YOUR VISION

You constantly generate mental movies and images based on your relationships, project experiences, and other events. No matter what the source of your mental images, there is one point that I want to drive home: You, and only you, are in control of your movies. Let's try a short experiment. Think about a glass of orange juice. Does that create a picture or image for you? I bet it does. Now, think about a horse. Can you see it? Change the color of the horse to pink. In a fraction of a second, you formed an image of the pink horse. Can you bring back the picture of the glass of orange juice? Of course, you can. You have control over the images that occupy your mind. However, when you don't consciously decide which pictures to play, your mind will look into the "archives" and keep replaying old movies on file in your mental library. Then, think about your vision as a project manager, analyze how things are happening in your projects, and try to visualize future images of desired outcomes. Think about how you would like to be treated by your team members and visualize the results. Indeed, you cannot change many situations, but you have the freedom to choose how you imagine them. Those mental images dramatically affect the future of projects and relationships with team members and other project stakeholders.

CHANGE THE MEANING OF THE OLD PICTURE

It does not serve you to deny what happened in an experience, no matter how painful or disappointing. You cannot, for instance, change the fact that the manager criticized you. You can, however, alter your interpretation of the event. When I told my story about my manager's criticism in front of my team, my perception was very negative. But I finally understood it some years later. At the time of original criticism, the meaning assigned to the experience is often "I am not good enough" or "My opinions are worthless." While this may be the interpretation of an inexperienced professional, the thought inadvertently gets carried into your life as a project manager. Today, though, you can consciously choose to view the situation differently. For example, the manager may have disagreed with you, but it was not a statement about your intelligence or overall worth. The manager may have been in a bad mood because of something else or may have had a point of view to share that eluded me at the time. That is the positive vision I am practicing now, seeing the opportunity to learn through the experience with that manager. *A powerful statement to enact is to say, I can choose to think differently about this situation.* I firmly believe that we often make incorrect interpretations of the words and actions coming from other people. Sometimes, it is because of our education, society, environment, or culture. That is much related to different people's motivational factors. According to Dr. Elias Porter, other people perceive their motivational values differently.

CREATE NEW PICTURES

We can create new mental pictures whenever we choose to do so. And when we develop (and concentrate on) new images that evoke powerful feelings and sensations, we will act in ways that support those new pictures. So, the first step is to create an image of your desired outcome. You are limited only by your imagination. Some project managers are terrified about public speaking. In survey after survey, it appears to be the first fear people have in organizations. So, when we ask people to consider making a speech, what kinds of pictures do they run through their minds? They see themselves standing nervously in front of the audience. Perhaps they have trouble remembering what they want to say. Run these images repeatedly on your mental screen, and you can be sure you won't succeed as a speaker. Instead, form a picture in your mind in which you confidently give your presentation. The audience members are listening to your every word. You look sharp. Your delivery is smooth. You tell a funny story, and the audience is laughing. In the end, you get a warm round of applause. People come up afterward to congratulate you. Do you see how these mental images can help you become a better speaker?

However, the pictures in your mind do not appear overnight. But, by being patient and persistently focusing on these mental images, you will automatically start acting in ways that support your vision. At the beginning of my professional career, I was involved in a project, working as a project leader by accident (without any training and knowledge). I had a team of six people, all older than me. My instinct told me I needed to listen to them before acting. Then, I started the project plan process with all of them. When the draft plan was ready, I delivered a presentation at the project kick-off meeting. It was my first presentation in my professional life in front of sixteen people. I was very nervous; I had butterflies in my stomach. But I listened to one of my oldest team members who told me: *Alfonso, be quiet; before starting your presentation, look at the people in the room and imagine they all are in pyjamas.* I put that small trick into practice, which considerably relaxed me. Immediately, I told myself: *Alfonso, you are the person who knows more about this particular project; go ahead and achieve a good result; you can do it.* The reader may think that it is a joke. However, it is true. That event was a key differentiator and critical professional learning in my life. Since then, I have used pictures and visualizations for various purposes, which work for me. Before presenting, I rehearse and rehearse, becoming more confident. Be prepared, and you will succeed.

PICTURE YOUR WAY TO SUCCESS

You must sell any product or service as a project manager or executive. You must see yourself as succeeding consistently. If you are not getting the desired results, there is no question that you are holding onto pictures of sales mediocrity or disappointment instead of sales success. Right now, think about your next meeting with a prospect. In your mind, how do you see the encounter? Are you confident and persuasive? Are you enthusiastically explaining the benefits of what you are offering? Is the prospect receptive and interested in what you are saying? Can you vividly see a successful outcome of your meeting?

Remember that you are the producer, director, scriptwriter, lighting coordinator, costume designer, and casting director of your mental movies. You get to choose how they turn out. You are paving the way for success in your sales career by mentally rehearsing and running successful outcomes through your mind. Of course, if you currently run images through your mind where the prospect rejects your ideas and has no interest in your presentation, you will attain minimal success from your sales efforts. You will attract those people and situations corresponding to your negative images.

RELAX AND INVOLVE YOUR SENSES

What is the best method to use when concentrating on your new images? Your mind is most receptive to Visualization when you are calm and not thinking about many things simultaneously. So, sit in a comfortable chair at home, close your eyes, and do deep breathing exercises to clear your mind and relax your body. Now, develop images that involve as many senses as you can. The more sights, sounds, smells, tastes, and touches you put in your pictures, the more influential the "pull" for you to make your vision a reality. Here is an example. Let's say you have always dreamed of owning a beachfront house in the Caribbean. Picture the white and peach-colored house. See the green palm trees slowly swaying in the gentle breeze. Smell the salt air. Feel the warm sand between your toes. Feel the sunshine on your face. Is this not paradise? And all this can be yours if you hold onto this image and do what it takes to achieve it. Also, remember that those images associated with strong emotions have even more power, so be sure to add positive feelings to your vision. For instance, when visualizing your ideal job, combine the vivid mental picture and the physical senses with the terrific emotions of pride and satisfaction you will have working in that new position.

Finally, do not be concerned with the quality of your images at the outset. Some people can create lively color pictures, while others have trouble getting more than a fuzzy image. It is also possible you may only be able to obtain a particular feeling at the beginning instead of a clear picture. In any case, do not worry about it. Do your best, and do not compare yourself to anyone else. There will always be someone better and others who are worse. Just be who you are. Your images will become sharper over time. The key is to spend several minutes each day running these new movies in your mind.

COMMIT YOURSELF

Commitment is an interaction dominated by obligations. These obligations may be mutual, self-imposed, explicitly stated, or may not be. You must distinguish between commitment as a member of an organization (such as a sporting team, a religion, or an employee) and a personal commitment, often a pledge or promise to oneself for personal growth. Formulating images of successful outcomes and running them through your mind is very powerful. But there is another technique you can use to accelerate your success. You can create visual aids to move you toward what you want. Let me share with you my story, another story of passion, persistence, and patience: In 1996, after presenting some papers written in English (which is not my native language), I realized that writing was enjoyable for me, and I set an objective of getting a book published during the next few years, at least in the Spanish language. I traveled a lot for project purposes over the next two years, but as soon as the number of trips slowed, I started working on my first book. And I got it. My first book

(*Dirección de proyectos—Una Nueva Visión*) was published in Mexico by Lito Grapo Editors, thanks to my good colleague and friend from Mexico, Rodolfo Ambriz, PMP. After getting that book published, and because of my good relationship with Randall L. Englund, PM Executive consultant, speaker, and author, I got an article published at PM Network from PMI. Randall helped me edit that article and encouraged me to continue writing. I followed his advice and set my new objective: to be a PM Network column contributor. And I did it one year later. Similar stories have happened over the last five years, and I have written this book for you. One of the tips I followed to achieve my objectives was to put that objective as an item in my daily tasks, as a reminder. That means to visualize the objectives. Make sure to look at your diary once a day and believe that you are moving toward that goal.

GETTING THE JOB YOU WANT

You can use visual reminders to your advantage in many ways; they are not limited to commitments. Here is an example of one of my colleagues, who we will call Francis. Francis applied as a candidate for the PMI Barcelona Chapter Board of Directors elections. While Francis stands an excellent chance of winning this election (and realizing his dream of becoming a president), he is still a little nervous, and doubts creep into his mind now and again. I suggested Francis make a handwritten sign where they will see it daily. I also recommended he write these words on a card he can carry in his wallet. By looking at those words throughout the day, Francis is conditioning his mind to view himself as a PMI Chapter Vice President. He will consider sitting at the Board of Directors room table in a meeting. As these images become stronger and stronger, Francis will take those actions that will bring this picture into reality. He will campaign more. He will ensure his party does everything possible to get the voters out on Election Day. While Francis could have formed solid mental images without using the sign, it has so much more power with the visual aid. The sign reminds Francis to think about being able to run as a Vice President and to run successful images through his mind. Of course, there are no guarantees that this will work for another person or will always work for you. But, once you try this for yourself, I think you will find that it is a potent aid to help you get what you want.

What project do you need to manage, and what do you like to achieve? Whatever it is, create a visual aid, and your mind will work to bring that picture into your life.

IT WORKS BOTH WAYS

Be very careful when using visual reminders. Some people use harmful aids with severe consequences. Bumper stickers offer a prime example. While riding in my car a few years ago, I noticed a bumper sticker on the vehicle in front of me. The bumper sticker read, "I feel sad, I feel sad, so off to work I go." I have repeatedly seen this same bumper sticker in the last few years. I am not kidding. There is nothing funny or harmless about this message. When you put something like that on your car, you are programming your mind to keep you sad. For example, every morning, Rose steps outside to greet the day and sees the statement, "I feel sad." When leaving work, she returns to her car and says, "I feel sad." This idea becomes embedded in her subconscious mind. She forms mental pictures associated with feeling sad. If you ask Rose why she lost that project deal, she will say she had bad luck. The truth is Rose is careless about what goes into her mind. The "harmless" little bumper sticker of today becomes your reality tomorrow. Rose is a perfect example of someone throwing more mud on an already dirty attitude window.

THE PLANNING PROCESSES

If you don't plan, it doesn't work. If you do plan, it doesn't work either. Why prepare a plan?

The nice thing about not planning is that failure comes as a complete surprise rather than being preceded by a period of worry and depression. The planning activities that you, with the help of your team members, will need to do for the project are listed below:

- To recruit and build the team
- To organize the project

- To identify and confirm the start and end dates through a project schedule
- To create the project budget
- To determine the customer requirements for the outcome
- To define the project scope boundaries—what it included and not included in the project
- To write a description of the outcome
- To decide who will do what
- To assign accountability

PLANNING FOR SUCCESS

A project must have a plan that demonstrates what is possible, shows an overall path and clear responsibilities, contains the details for estimating the people, money, time, equipment, and materials necessary to get the job done, and will be used to measure progress during the project and act as an early warning system to be successful. Planning for success is also based on team member's commitment. An example of responsibility was the "Manhattan Project": It was, among other things, a gigantic industrial and engineering construction effort run by the military under great secrecy, rapidly accomplished, using unorthodox means, and dealing with uncertain technologies. Its central purpose was to develop and build an atomic bomb as quickly as possible to end the war. It got underway in June 1942, but only with the appointment of Army Corps of Engineers officer Colonel (quickly promoted to Brigadier General) Leslie R. Groves on September 17, 1942, did it become an all-out crash program. In less than three years, they tested the first bomb on July 16, 1945, in the New Mexico desert. Three weeks later, on August 6, the Japanese city of Hiroshima was bombed, followed by the bombing of Nagasaki on August 9. The war was over five days later, on August 14.

One of the characteristics of the Manhattan Project was the unconventional practice of conducting the *research, development, and production phases simultaneously* rather than following a step-by-step sequential path, which is the standard and slower way. Because every minute counted in wartime, all the steps were compressed and done in parallel rather than one after the other. Occasionally, there was a problem, but we corrected the wrong choice. With work just about completed, General Groves decided that an alternative design would be better. And so they stripped the just-installed machinery out of the plant and put new machines in their place. This practice of compressing the different stages does not always work and should be a cautionary tale for any future projects. Total program authority was vested in Groves. He had the complete support of the president and the other high officials of the administration. The complete resources of the U.S. Treasury were available to him. Ultimately, the project cost about $2 billion, about $30 billion in today's dollars. The objective was clear, unmistakable, finite, and well-defined. Compartmentalization and maintaining security kept people focused on their assignments and responsibilities to achieve them. Each element had its task, and all were carefully allocated, assigned, and supervised so that the sum of the parts resulted in accomplishing the mission. Command channels were clear-cut, well-understood, and direct. Authority delegated with responsibility. Large staff were avoided, especially in the Groves Washington office. Higher-level people knew one another from past experiences and could quickly communicate to solve problems and make decisions. They kept a minimum of written communication. Most business was done verbally by phone or face-to-face. Groves did not make decisions based on staff studies, committee reports, consultant written opinions, or the like.

The Manhattan Project was administered very much according to the organizational model and practices of the Army Corps of Engineers. It is unsurprising since Groves, the purest of specimens, was its head. The model emphasized decentralization but through clear lines of command to the top. I think the Manhattan Project's success lies within the culture and organization of the Army Corps of Engineers and the high quality of the officers that ran it. They used them for big projects. Size did not faze them. Groves, in his earlier capacity, just before being selected Manhattan chief, oversaw more than eight billion dollars' worth of domestic army construction projects during the mobilization period from 1940 to 1942. Groves always projected an optimistic attitude, which inspired others. They could sustain their morale only if everyone thought it possible. If Groves showed doubt, hesitation, or fear, it might infect the others and undermine the project. Groves typically set completion dates that he was sure could not be met. Keep the bar high, and people will work harder to jump over it. He set completion dates, which he considered impossible.

Only in this way could Groves be confident that every effort would happen, and no one could think of easing up if he had too easy a schedule. Success is not a matter of luck, Groves said, but the result of mental and physical capacity, endeavor, determination, and, in considerable measure, competent management.

Examining the Manhattan Project and such collaborative efforts, we conclude:

- To achieve success, start with superb and gifted people.
- It is best that they produce something tangible as opposed to working on an abstraction or an idea.
- Young people usually are more energetic, confident, and curious and, thus, are more likely to work harder and longer.
- It is all the better if the moral purpose drives the undertaking. Put this particular population in an isolated spot without any distractions. Living in Spartan conditions makes work the focus, with no distractions.
- This tendency to escape into the work may result in ignoring or not having the time to reflect on the final result.
- The cooperation of the many parts toward realizing the overall goal is essential. Ensure that those below have faith in their leaders and that the leaders have faith in those below.

In conclusion, while learning and applying history lessons is essential, we must remain cautious about making too easy analogies.

CASE STUDY

In 1992, I delivered my first project management presentation at a multinational company conference in San Jose, California. I decided to submit a paper at a conference organized by the Project Management Initiative of HP. It was challenging because my English level was low. However, I had a clear idea: to share my project experience with my colleagues at HP internationally. Then, I submitted my paper. Initially, I got some comments and requests from the organizers and did my best. Finally, my paper was accepted. I had three or four months to prepare my presentation, and

I prepared my project plan as a project. I found some HP Spanish colleagues who had worked in the US and asked them for advice to review my plan and manage the project risks. I went to the HP library and looked for past HP Conference proceedings, trying to familiarize myself with paper styles. And finally, I joined an English class every day. During the first two months, my English classes focused on grammar and sentence construction, and during the last month, I rehearsed my presentation every day. My English teacher recorded me with a video camera, and we reviewed my expressions and sentences together (project snapshot). My first feeling was that watching myself in the video and listening to my voice was challenging. I did not feel good. However, I believe this was an excellent method to receive feedback and make improvements. I was encouraged by a Spanish colleague who lives in the US, and that reinforced my positive behavior.

I remember the faces of HP colleagues like Tom Kendrick, PMP from HP, and others in that room. Initially, I was very nervous but felt much better after some minutes. I remembered how I practiced so many times to achieve my objective—delivering a presentation in English and being understood by the audience at that specific event. I visualized my scenario many times and believed I could do it. And I did; my talk was successful. Almost everyone understood me because I rehearsed many times in front of a mirror and watched the videos. I was pretty surprised when I saw my presentation recorded by a colleague. My improvement through that long process was huge. My English was not perfect, but my delivery was good, showing my preparation and knowledge about the subject. So, I was credible and got some applause. I believe that you can plan for your success, and if you are passionate, persistent, and patient, you can achieve good results. This example may not appear to be a big success for some readers. It was not, but it was a first step to convince myself I could accomplish my goal through planning, practice, and courage.

SUMMARY

I describe Visualization as "movies and pictures of the mind," "inner pictures," or "images." We all store pictures in our minds about the type of relationships we want, the degree of success we will attain at work, the extent of our leadership ability, and so on. I can summarize the ideas and thoughts explained in this chapter as follows:

- I firmly believe that many times, we incorrectly interpret the words and actions coming from other people, perhaps because of education, society, or culture.
- I strongly related to people's motivational factors.
- Public speaking provides opportunities to change attitudes through revisualization and achieving a positive outcome.
- You, as a project manager or as an executive, are involved in selling any product or service.
- It is vital to see yourself succeeding consistently.
- If you are not getting the desired results, you probably hold onto pictures of sales mediocrity or disappointment instead of sales success.
- Be very careful when using visual reminders. Some people use harmful aids with severe consequences.
- Formulating images of successful outcomes and running them through your mind is very powerful.
- What project do you need to manage, and what do you want to achieve? Whatever it is, create a visual aid, and your mind will work to bring that picture into your life.

TOOL—MY PERSONAL VISION TEMPLATE

Spouse

(Describe what your vision is about your wife or your partner.)

Example: To be a good husband and help my wife to walk together to our future.

Children

(Describe your children's vision.)

Example: To be a reference as a parent supporting them to grow personally in a climate of love and respect in our family.

Career

(Describe your career vision.)

Example: To be a professional reference point to other professionals in the project management field.

Moral

(Describe your moral/ethics vision.)

Example: To be a supporter for others who need some help financially, emotionally, and physically.

Extended Family

(Describe your extended family vision.)

Example: To be of help to the extended family to understand them better.

Prof. Involvement

(Describe your professional involvement vision.)

Example: To continue serving the PMI community as a volunteer.

Spiritual

(Describe your spiritual vision.)

Example: To be a better person, giving more to my family and friends.

Physical

(Describe your physical health vision.)

Example: To be healthy, taking care of my mind and body health.

Community

(Describe your community contribution vision.)

Example: To expand the benefits of the project management profession to my local community.

Financial

(Describe your financial vision.)

Example: To keep financial health during the next year.

Recreational

(Describe your recreational vision.)

Example: To spend more free time with my spouse and family, taking at list one week of vacation to spend free time together.

Action Plan

(Describe your action plan for the next six months. Please review it every year.)

4

Make a Commitment

Courage is doing what you're afraid to do.
There can be no courage unless you're scared.

Eddy Rickenbacker, US WWI aviator & businessman
(1890–1973)

The Merriam-Webster dictionary defines courage as the mental or moral strength to venture, persevere, and withstand danger, fear, or difficulty. Courage, also known as bravery, will, intrepidity, and fortitude, is the ability to confront fear, pain, risk/danger, uncertainty, or intimidation. After inventing the light bulb, journalists asked Thomas Edison where he drew inspiration. He said: "*I find out what the world needs, then I proceed to develop.*" The key to getting what you want is the willingness to do whatever it takes to accomplish your objective. Now, before your mind jumps to conclusions, let me clarify that in saying "*whatever it takes,*" I exclude all actions which are illegal, unethical, or which harm other people. So, precisely what do I mean by this "willingness"? It is a mental attitude that means that if it takes five steps to reach my goal, I'll take those five steps, but if it takes fifty-five steps to reach my goal, I'll take those fifty-five steps, and so on. We usually do not know how many steps will be required to reach our goal, but this does not matter. To succeed, all that's necessary is to commit to doing whatever it takes, regardless of the number of steps involved.

Where does persistence fit in? Persistent action follows *commitment*. It would help if you were committed to something before you persist in achieving it. Once you have committed to achieving your goal, you will follow through with relentless determination and action until you attain the desired result. When you save and are willing to do whatever it takes, you attract the people and circumstances necessary to accomplish your goal. For instance, once you devote yourself to becoming a better project manager, you might

DOI: 10.1201/9781003485674-4

suddenly "bump into" a new professional resource or "discover" a forum on this topic. It is not as if these resources never existed before. It is just that your mind never focused on finding them. Once you commit yourself to something, you create a mental picture of what it would be like to achieve it. Then, your mind immediately goes to work, like a magnet, attracting events and circumstances to help bring your picture into reality. However, it is essential to realize that this is not an overnight process; you must be active and seize the opportunities as they appear. William Hutchison Murray said: "Until the commitment of one person, there is hesitancy, the chance to draw back, always ineffectiveness." Concerning all acts of initiative, there is one elementary truth, the ignorance of which kills countless ideas and splendid plans, that the moment one commits oneself, Providence moves, too.

> All sorts of things occur to help one that would otherwise never have oc-curred. A whole stream of events issues from the decision, raising in one's favor all manner of unforeseen incidents and meetings and material assis-tance, which no man could have dreamed would have come his way.

You do not have to know how to achieve your goal at the outset. When you are willing to do whatever, it takes, the proper steps are often sud-denly revealed to you. You probably will meet people you could never have planned to meet. Doors will unexpectedly open for you. It might seem like luck, or good fortune is smiling on you; in truth, you will have created these positive events by committing and, thus, instructing your mind to look for them. Let me provide an example. In July 2008, I launched a new idea to the market, the Project Portfolio Event. It was a one-day-long pro-fessional event on project portfolio management.

I was committed to making that event happen. I had a positive attitude in dealing with that project. However, everything was not rosy in my path. Life tested me to see how serious I was about achieving my aim (to make it hap-pen). I found many obstacles: Customer registrations were only a few at the beginning, Sponsors did not pay us early, and the worldwide market crisis seriously affected us. I made some mistakes and suffered disappointments and setbacks, some of which may be severe and even tempt me to abandon my goal. For instance, I had five people on my team, and although I always encouraged them, registrations were still low when we only had three weeks left for the event. But I did not give up; I talked to them and asked them to move forward. I continued contacting potential attendees and doubled the number of attendees two weeks before the event. That's when it becomes essential to follow the sage wisdom of Winston Churchill, who said: "Never,

never, never give up." If you have committed to accomplish a goal, you can overcome temporary defeats, and you will triumph. Genuine commitment inspires and attracts people. It shows them that you have conviction.

REFUSE TO QUIT

I have learned a lot about the magic of commitment from my ex-colleague and friend John P. John developed his project management career at a computer manufacturer organization in the UK. From a very young age, he liked project management unconsciously. John worked as a project manager at a multinational computer manufacturing company some years later. However, he had a straightforward project in his mind (getting retired around his fifties). Then, his goal was to save money through hard work doing his preferred work (project management). He managed many projects during his career and always tried to take on significant challenges. He traveled a lot, and sometimes he felt exhausted. However, he never quit. He continued and managed his last program as program manager of the Y2K (year 2000) project for a multinational company. He is now in his 70s and happily retired. He is enjoying himself managing the project of building his new house. He is an example of persistence. We are talking about commitment, but we are also talking about perseverance and patience, about keeping a good attitude in the face of rejection.

Another story is Michael, who lives in Madrid, Spain. At forty five, Michael graduated from a Spanish Project Management school. Then, he started to prepare to take a PMP exam and become certified. On his first attempt at the examination, Michael failed. On his second attempt, he failed, too. Michael tried the third time, the fourth time, the fifth time, the sixth time, and the seventh time. His main problem was that he took the exam in English, and English is not his most vital skill. Most people would have quit, but not Michael. He was persistent; he took the exam for the eighth time and passed. Michael is a clear example of persistence, a positive attitude, and courage. He is managing and consulting projects and is happy doing that job. Refusing to quit is to show your courage. I remember a quote from Miguel de Cervantes (Spanish author): *If you lose your wealth, you lose a lot; if you lose a friend, you lose much more, but if you lose your courage, you lose everything.* See Figure 4.1.

A good project manager must empower, inspire, and motivate their team members to conquer the project's daily obstacles and issues. Projects always generate problems. Work to solve them.

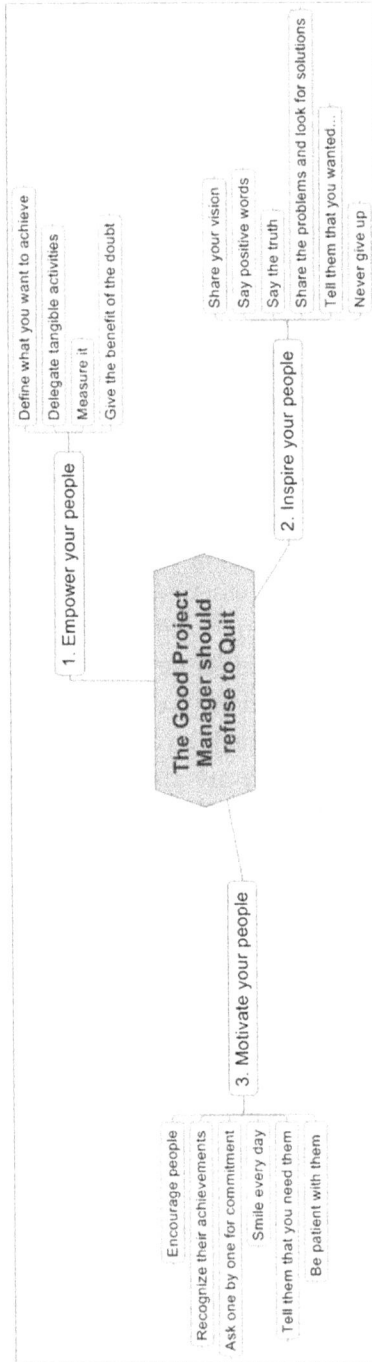

FIGURE 4.1

The good project manager should refuse to quit.

TIME TO MAKE A COMMITMENT

Now, let's assume you have a goal in mind. The next question is, "Am I willing to do whatever it takes to achieve this goal?" If your answer is, I'll do just about anything, except that I won't do this and that…, then honestly, you are not committed. Without commitment, you will derail and not achieve your objective. For instance, many people start a new business with this approach: "I'll give it six months to prosper. If things don't work out after six months, I'll quit." That mental attitude does not lead to success. When I left Hewlett-Packard (HP) in 2002 to start up my own business (a new project), many people from HP told me: "Alfonso, you are a fool; business life is challenging outside HP." However, I made my commitment to creating my business. Now, several years later, I am proud of being successful with my business. I passed over difficulties, people problems, customer issues, and financial challenges, but I got it, and I hope to continue managing my business for many years. Now, I am not suggesting that you bow ahead without a plan and hope for the best. Of course, you should set timetables, deadlines, and budgets, so you stay on course and succeed as quickly as possible. But the reality is, despite your most careful plans, you don't know how long it will take to achieve your goal, and you cannot foresee all the obstacles that will cross your path.

That's where commitment separates the winners from the losers. The committed people are going to hang in and prevail no matter what. And if it takes a little longer than they thought, uncommitted people will give up the ship when things don't go their way. Now that you have learned about the power of commitment, it is time to apply the principle. So, go ahead. Select a goal you have a burning desire to achieve. Commit to do whatever it takes to achieve this goal. Start moving forward and get ready to notice and take advantage of all opportunities that come your way. Then, follow through with persistent action and get prepared to succeed. How do you improve your commitment? Measure it, know what's worth dying for, and make your plans public. I had to develop new products and services when I created my company. I often prepared a draft of the product description, content, objectives, and audience. I announced the availability date for that service or product to my customers and peers. That approach worked for me; it generated a strong commitment within me and with my people.

COMMITMENT AND PROJECT SUCCESS

When assigned as a project manager to the General Treasury of Spanish Social Security in Spain, I never imagined what I learned about the power of commitment from that project. Let me share my experience with you:

The Background:

HP and the Data Collection Center of the General Treasury of the Spanish Social Security achieved one of the most advanced solutions for electronic document management. Using an image management system, they devised a solution to digitize, store, and process the Spanish Social Security contribution forms (TC1, TC2, etc.) despite the tremendous volume of documents processed—about two million pages per month. CENDAR, the Data Collection Control Center, is a Spanish public entity within the General Treasury for Social Security. It acts as a control center for information relating to collecting contributions made by employees and companies about Social Security.

The Situation:

The General Treasury of Social Security used a manual data entry system to input information regarding Social Security contributions. Because of the substantial volume of paper generated by this operation, it became necessary to design a system allowing electronic storage of the forms for subsequent processing. In July 1991, the criteria for a public tender to purchase an information systems solution for the Electronic Data Collection Exchange Center (CENDAR) were published. HP won the proposal.

The Solution:

Good cooperation between the customer and HP in the CENDAR project team made it possible to make a clear definition and its objectives. The teamwork between the customer and the HP team and their commitment were the fundamental reasons for HP being able to meet the client's requirements on time, on cost, and with the expected quality level. I spent long journeys at the beginning of the project to create a good team. The team was formed by young people (from twenty-five to thirty years old), but they were anxious and ready to learn. The cooperation between CENDAR professionals and

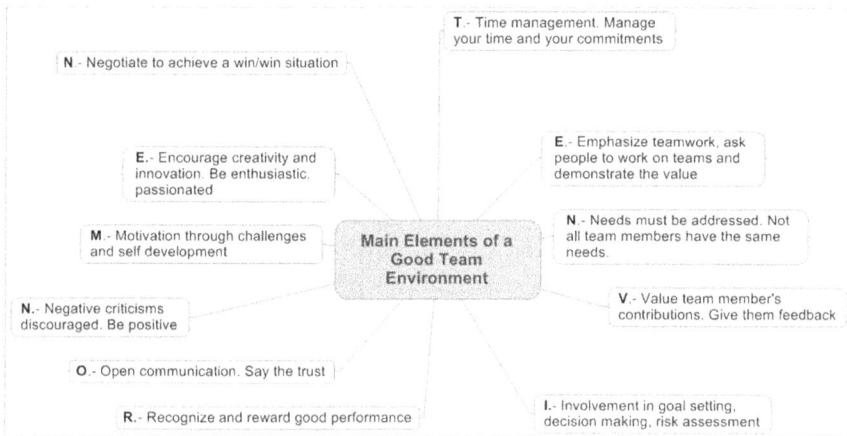

FIGURE 4.2
Main elements of a good team environment.

HP during the solution design and implementation phases gave the customer a state-of-the-art document management system. This real case showed that total commitment concerning the client's requirements and prime responsibility for the whole project were the keys to success in developing the new system. The team I created and managed in this project taught me that together, everyone achieves more. During this project, I made a suitable environment; it was not easy because my team had people from different organizations and with different cultures and expectations. However, you can see the main elements I consider for a good team environment in Figure 4.2.

CASE STUDY 1

When I finished my Computer Science Engineering degree in 1984, I started my Ph.D. studies, but I joined Secoinsa firm as my first job. I began to travel frequently, so I interrupted my Ph.D. studies some months later. However, I promised myself that someday I would finish my Ph.D. Although many of my professional colleagues told me: "you do not need a Ph.D. now," I always thought if I pursued discovering something new, I would learn something. Twenty-five years later, at forty-nine, I committed to achieve my Ph.D. in Project Management. I did not find Universities offering that specialty in Madrid for the weekends, so I decided to do it at Zaragoza University. Zaragoza is about three hours by train. But that was not a big deal for me. Since the beginning of the program, the Program Director has transmitted negativism to all his students. I was very disappointed about his behavior because it is not a good skill to show up in a project manager practitioner. It impacted me very negatively. I almost decided to quit, but I was persistent; I was patient and determined to keep asking myself: What do you want to achieve, Alfonso? I attended all classes regularly during the first year to accomplish my objective. Unfortunately, I could not finish some homework for business reasons and failed three of the nine program subjects. But I did not quit. The following year, I passed. In the third year, I worked on my pre-thesis and presented my thesis project. To stay focused on that purpose, I put my thesis project activities on my daily agenda to advance step by step. I am a frequent business traveler, and it has sometimes been difficult. Then, I got the DEA (Master Degree in Project Management Certificate), a pre-requisite to present the Thesis. I traveled to Zaragoza once per month to meet my mentor, who always told me: *"Alfonso, you will not be able to achieve it because you have your own business; you must work for your customers to survive. You are managing a consulting firm, and it is not your priority."*

Some not polite answers came to my mind. I always told my mentor I wanted to continue and would be very persistent. He continued with the same talk: "Alfonso, you will not…, you will not be able to…The last time, I was furious." I treated him to lunch and told him: "I love you too." I talked to him clearly, explaining that I could survive because of my passion, persistence, and patience, even when some of my colleagues quit. Our relationship became worse, and I decided to leave that University. He demotivated me. Two years later, I continued with another University in the Basque Country in Spain but was unsuccessful. Managing a project, traveling weekly, and doing a Ph.D. are incompatible. Although I decided to stop my Ph.D. project for a while, in my mind, I had one pending task to accomplish: getting my Ph.D. finished. Finally, in 2019, I enrolled in a University in the North of Europe (ISM) where, after five years of part-time and great supervisors, I finished my Ph.D. successfully.

CASE STUDY 2

After being a PMI member since 1993, and contributed to several Congress by submitting some papers and presentations, being one of the founders of the PMI Madrid Chapter and trying to belong the PMI Madrid Chapter

Board of Directors, I made the commitment of starting up a PMI Chapter in Barcelona. It was not easy because my residence was in Madrid but using my passion to move forward. First of all, I searched for some allies and support among my project management colleagues and we submitted a proposal to PMI. After almost one year, the PMI Barcelona Chapter was chartered with the first twenty-five members.

Every fifteen days, my colleague J.S. traveled with me by car early in the morning and drove to Barcelona to run a Chapter Board meeting, doing the meeting, having lunch together, and driving back to Madrid after lunch. I used my passion for the profession to engage a great team of project management believers who followed me. At the beginning, it was not a path of roses, my colleague J.S. and I were not from Catalunya and it was challenging to gather and complete all the administrative staff. J.S. did a great job being supported by the rest of team members who were from Barcelona.

A great challenge was to make our first chapter event happen. We made it and got close to one hundred people and international speakers from the PMI community. Then we got organized frequent members meetings where professionals share their project management experiences. We searched for PMI Chapter Sponsors and we started with Microsoft, who is still supporting that PMI Chapter. Several PMI Barcelona Chapter presidents followed my work over the years and contributed to its membership and services grow. Now in 2024 was the twentieth PMI Barcelona Chapter anniversary, and I am proud of this chapter's growth, success, and future progression.

SUMMARY

The key to getting what you want is the willingness to do whatever it takes to accomplish your objective. Here are reminders of essential best practices from this chapter:

- When you commit and are willing to do whatever it takes, you attract the people and circumstances necessary to accomplish your goal.
- Commitment is necessary to be an effective leader; genuine commitment inspires and attracts people. It shows them that you have conviction.
- A good project manager must empower, inspire, and motivate team members to conquer daily obstacles and issues. Projects always generate problems; work to solve them.

- How do you improve your commitment? Measure it, know what's worth dying for, and make your plans public.
- Always hang on to your passion, persistence, and patience, not only in your projects but also in your personal life.

TOOL—COMMITMENT ASSESSMENT

Select the number closest to representing how true the statement is for you right now. You will obtain a final score once you have answered all the questions. This test helps assess how committed you are to yourself.

Question	Scoring: from 1 to 4 (from strongly disagree (1) to strongly agree (4))
1. Do you dedicate time to yourself each day?	
2. Do you think "selfish" is a negative word?	
3. Do you feel guilty saying NO to your children, partner/spouse, family, and friends?	
4. Does your job take priority over you and your family's needs?	
5. Do you make daily choices based on your priorities being honored first?	
6. Do you have a list of things you wish to do or do?	
7. Do you know how to make yourself a priority?	
8. Do you fear disappointing yourself?	
9. Are you ready to say YES to yourself?	
10. Do you believe making yourself happy first allows you to make others happy?	

Scoring Results

Filling in this table will help you be more conscious about your commitment and will allow you to prepare a plan to improve. Monitor your progress, and you will succeed.

5

Convert Your Project Issues into Opportunities

Every adversity carries with the seed of an equivalent or more significant benefit.

–Napoleón Hill

What is your immediate reaction when faced with project problems or setbacks? If you are like most people, your first impulse is to complain. "Why did this have to happen to me? What am I going to do now? What a disaster!" This response is only natural. However, you have a choice after the initial disappointment disappears. You can either wallow in misery and dwell on the negative aspects of your project situation or find the benefit or lesson the problem offers. Yes, you will probably face a period of uncertainty or struggle, but there's always a flip side to the difficulty. A "problem" is often not a problem at all. It may be an opportunity. For instance, a problem may point out an adjustment you can make to improve certain conditions in your project. Without the problem, you never would have taken this positive action. What started as an adversity ended up as a golden opportunity. My colleague and friend Randall Englund compelled himself to find a new position during a low point in his career. After an arduous search, he signed on to a project management initiative. That new position launched his career as a project management consultant. How about when you know that a particular job is perfect for you; you had a great interview and could not wait for the offer? But the offer never came—someone else got the job. You were devastated. Days or weeks later, a new job came along, and you realized that the first position was much less desirable than the one that came along later. The earlier rejection was,

DOI: 10.1201/9781003485674-5

in fact, a blessing. Another example is the deal on the "dream house," which falls through only to be replaced by something even better.

THE BENEFIT

Human beings and professionals come to appreciate the important things in life when we live through difficult times. Let me share a project I managed in the North of Spain: *I was a project manager for an infrastructure customer project. I had two people from HP assigned to my team, but I also had a significant subcontractor as part of the team. That subcontractor had previous business relationships with my customers, so they were very confident in each other. I was the youngest guy among all the main project stakeholders, so I experienced much stress at the beginning of the project. I was in an uncomfortable environment. I was the project manager and the primary person responsible for the project. This project was critical for HP, so I talked directly to my project Sponsor (the HP Managing Director). I asked him for advice and support. He encouraged me to ask for help whenever I needed it. Then, I tried to establish a closer relationship with the customer, talking to him, treating him to lunch, and providing him with technical explanations and project status. Little by little, I gained credibility in his eyes. The subcontractor observed my behavior and treated me to having lunch together. His approach became more open and sincere than at the beginning of the project. That achievement took more than six months, but I got it. My fear in front of the subcontractor disappeared, and I demonstrated, first of all to me and then to others, that I could convert an issue into a benefit to facilitate project success. See Figure 5.1.*

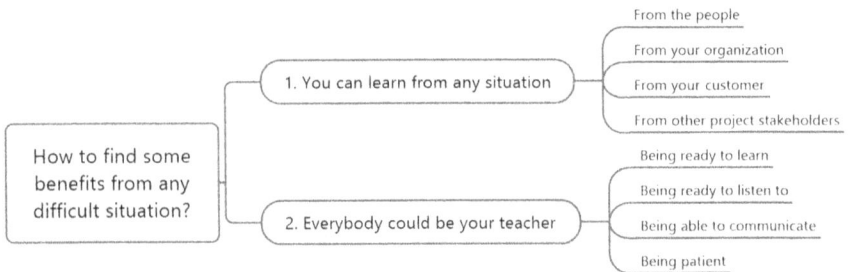

FIGURE 5.1
How to find some benefits from any problematic situation?

I firmly believe that you can find some benefit from any problematic situation. You can learn from any situation and anybody if you think everybody could be your teacher. You, the reader, could be my teacher.

FROM PROJECT FAILURE TO PROJECT SUCCESS

The road to success often travels through adversity. I was the project manager for a ten-million Euro customer project in the Spanish General Treasury of the Social Security. It was a software development and infrastructure project. Three months after the project started, the division that supported the software product we used closed and fired all their employees without any advice. I felt highly stressed in front of the customer. There was a high probability that the project would fail. At the same time, one core project team member left the company. He was the most knowledgeable professional about crucial technical aspects of the project. However, the company supported me and invested money to find a subcontractor to provide us with an alternate software package. The company empowered and encouraged me to establish relationships with subcontractors. On one hand, I needed to find another professional with sound technical knowledge. It was an excellent opportunity for me to deal with people from different organizations with different cultures and visions and negotiate with them. At the same time, I had to interview various candidates to substitute for my project core team member. I learned a lot from that situation, personally and professionally, and one year later, the project finished successfully. The customer was happy using the system, and my managers valued my effort and improvement.

MY PROJECT MANAGEMENT CAREER

My professional career transition is another example of how benefits come from problems and difficulties. Adversity brings out our hidden potential. Since 1985, I wanted to be a project manager. I loved dealing with people and enjoyed managing issues and difficulties, but I was probably unaware. My first job was for a Spanish computer hardware manufacturer called Secoinsa in 1984. My job was as a technical software product specialist.

After one year of experience, I had the opportunity to manage my first project with the "Banco Hispano Americano." I worked on that project by accident—without any project management experience. I had a team of four professionals, all older than me, but I was patient and perseverant, trying to learn from them. It was my first time telling my team members, "I need you." It worked so well. I ran into many difficulties with that one-year-long project. I had to deal with people, customers, and technical issues, but I survived. I learned to use common sense more and more. I became conscious of how difficult a project management discipline can be. On the other hand, I verified that it was not impossible.

One year later, I moved to another company, Digital Equipment Corporation. I started in the presales department as a technical consultant, but after a few months, I was assigned to different projects, first as a team member and some years later as a technical team leader. I had the opportunity to receive basic project management training, and I discovered I liked it. After four and a half years, I moved to ICL, another computer manufacturer, as a project manager. I started acting as a project manager for ICL projects in Madrid and Valencia. I developed leadership skills there and created good relationships with customers. However, the company situation and the lack of management support encouraged me to move to my next job in a different organization. I proceeded to HP Spain in the Madrid office. I continued working as a project manager at HP for almost fourteen years. It was a personal and professional challenge. I had the opportunity to learn and manage complex projects. I had an excellent opportunity to learn from my managers and project manager colleagues. I helped to start up three project offices. I discovered the existence of professional project management associations, and I joined the Project Management Institute (PMI). A vast window opened when I attended my first International Project Management Congress. Project management became my passion after my first attendance at a PM Congress.

I discovered how much I had to learn and that the profession greatly motivated me. I took care of my professional network year by year. I published my first book, *"La Dirección de proyectos una Nueva Visión,"* published in Mexico with the help of my friend and project professional Rodolfo Ambriz. At the same time, I started teaching project management in Business Schools in Málaga and Madrid. It was an extraordinary effort and a way of feeling more professionally recognized. I left HP to start my own business. It was a challenging project, but after a few years I was happy and could say "Today is a Great Day."

I had a team working with me that also believed in my principles: Passion, Persistence, and Patience. During those seven years, I had financial, people, and organizational problems, but I always had a positive attitude to learn from successes and failures. I always had the strong support of my wife, Rose, who believes in me. She taught me a lot about dealing with difficult people. I continued writing articles and submitting papers to Project Management Congresses. I wrote two books with my good friend Randall L. Englund and three more books alone. I like writing and always try to contribute to the project management profession through articles, presentations, and speeches. I continue discovering how much I can learn from my customers, students, and colleagues in the project management field. The expansion of the project management profession is a great challenge. The project management career never will end for me. The problems and issues when managing projects always represent opportunities to learn.

PROJECT ADVERSITY

Let's examine some ways adversity can serve you. Adversity provides you with perspective. Once you have recovered from a life-threatening illness, a flat tire or leaky roof no longer seems so troubling. You can rise above the petty annoyances of daily living and focus attention on the essential things in your life. Adversity teaches you to be grateful. You develop a deeper appreciation for many aspects of your life through problems and difficulties, especially those that involve loss or deprivation. It is trite but true that you do not usually appreciate something until somebody takes it away from you. When you have no hot water, you suddenly value hot water. Not until you are sick do you cherish good health. The wise person continues to dwell on blessings, even after the loss or deprivation has passed. Remember, you are constantly moving toward our dominant thoughts; therefore, concentrating on what you have to be grateful for brings better things into your life.

Adversity brings out your hidden potential. You emerge emotionally stronger after surviving a problematic deal or overcoming an obstacle. Life has tested you, and you were equal to the task. Then, when the next hurdle appears, you are better equipped to handle it. Problems and challenges bring out the best within you; you discover abilities you never knew you possessed. You would never have found these talents if life had not made you travel over some bumpy ground. Adversity reveals your strengths and capacities and beckons you to develop those qualities further. For example, I managed *a project for a Savings Bank in Spain. The project was an Infrastructure project plus a Change project. I was the project manager, and I had a team leader reporting to me from a consulting company to lead all the process changes, job changes, and organizational communication. After three months of work, the customer complained about the Change part and asked me to fire the team leader. I tried to convince my customer that it would be too risky, but finally, because of political reasons, the team leader from the consulting company was out of the project. I had to assume all the responsibilities. It was tough, but I was honest with my manager and customer. I learned a lot. I got support from my customers daily, and I achieved good results. My effort was worth it.* Adversity encourages you to make changes and take action. Most people cling to old, familiar patterns regardless of how boring or painful their lives have become. It often takes a crisis or a series of difficulties to motivate them to make adjustments. Problems are frequently life's way of letting you know that you are off course and need to take corrective action. See Figure 5.2.

Adversity teaches valuable lessons. Take the example of a failed business venture: The entrepreneur may learn something that enables them to succeed spectacularly in the next venture. Adversity opens a new door. A relationship terminates, and you go on to a more satisfying relationship. You lose your job and find a better one. In these instances, the "problem" is not a problem but rather an opportunity in disguise. One door in your life shut, but a better one awaits. Adversity builds confidence and self-esteem. You feel competent and gain confidence when you muster your courage

FIGURE 5.2
Adversity.

and determination to overcome an obstacle. You have a more incredible feeling of self-worth, and you carry these positive feelings into subsequent activities.

LOOK FOR THE POSITIVE

Sure, you will have your share of problems and adversities in life. I am not suggesting that you deny your emotions or refuse to face reality when tragedy strikes. What I am saying is, do not immediately judge your situation as a tragedy and dwell on how bad off you are. Sometimes, you cannot instantly spot the benefit of being in your situation, but it does exist. You always have a choice. You can view your problems as unfavorable and become gloomy and depressed about them. Let me assure you this approach will only make things worse. Or, you can see every seemingly negative experience as an opportunity, something you can learn from and grow from. Believe it or not, your problems are there to serve you, not to destroy you. So, the next time you suffer a problem or setback in your life or project, don't get discouraged or give up. Do not let problems cloud your attitude window forever. Clear off that cloudy window. After the dust settles, you may find that you can see better than you did before.

Just remember the words of Napoleon Hill: "Every adversity carries with it the seed of an equivalent or greater benefit." For example, *two years ago, I started a new personal project: to graduate in leadership from the Leadership Institute from PMI (Project Management Institute). PMI admitted me to the class. The training program had two possibilities: enrolling in the US or the EMEA (Europe, Middle East, and Africa) class. I participated in the first "face-to-face" session organized for professionals coming from EMEA. Four months later, PMI scheduled a second face-to-face session, but in the meantime, my mother suffered an accident (she broke her leg), and she had a surgery intervention. That accident happened just one day before PMI scheduled their second face-to-face EMEA session. Then I had to make a decision (go or not go). My decision was not to go and to stay with my mother at the hospital until she recovered. Initially, I was frustrated; much effort, homework, study exercises, and reading appeared lost. I could not attend the second training session.*

The Training Program director from the Leadership Institute called and suggested that I enroll in the second class for the US and attend their second "face-to-face" training session. So, I did. Then, I received the third training session and completed the whole program for the rest of the year. Because of that adversity, I now know double the number of professional colleagues worldwide. I had the tremendous opportunity to meet all my colleagues from my first group in Europe and a second group of people in the US. It was great to increase my network and knowledge after a personal crisis. Ask yourself what you have learned from your trying experience and focus on moving forward and growing. In times of crisis, always strive to maintain an optimistic attitude and an open mind, for this environment will allow you to find the benefit in your difficulty.

CASE STUDY 1

I was the project manager for a project in the South of Spain. I lived outside the home from Monday to Saturday every week for over three years. The project was critical for the customer, and we could not fail. This project generated a lot of stress and project and personal issues. I had a large project team—eight people from my company and more than thirty people subcontracted from different organizations. After some months, we achieved our first project milestone, but afterward, people started to feel depressed and demotivated. I analyzed the situation and understood that we worked more than twelve hours daily. We were not efficient. I led by example. I met the whole team and told them I would leave the work at 6:30 pm each day, and I did it. The result was awe-inspiring; people followed me and acted similarly. The project productivity increased dramatically.

On the other hand, I also found people on Monday morning telling me: *Oh Alfonso, it is Monday, my God! A long and sad week.* I immediately told them: *Don't worry, now it is 9:30 am, and we will have a coffee break in one hour. After the break, lunch will be coming soon; after lunch, it is almost Tuesday.* I always tried to be positive with my people and transmit confidence and passion to complete our job. It was beneficial for us during all those three years. Over the years, I have discovered that professionals, as human beings, develop special skills in difficult and stressful situations.

CASE STUDY 2

I managed a project for a Spanish company in the distribution sector in Barcelona. My direct contact was the Operations Manager of the company, and the headquarters was in Barcelona, having distribution terminals in Bilbao, Tarragona, and Valencia. The project involved moving that company from a functional organization to a project-based one, taking into account that the main organization business were operations. From the very beginning, I worked with the Board of Directors to explain the transformation project in detail and get their buy-in. I insisted in the need of management commitment and support for the project success. Not all the Board members liked this project. The company was obtaining good results, and now somebody from outside came to bother them, and they scared about the daily impact from that project. On the other hand, most components of the Board of Directors in that company had no idea about project management, but all of them thought there was some room for improvement in their company contributing to some process's efficiency. I ran into some problems with some managers who could not understand the added value of project management and were not inclined to collaborate at the beginning. However, my main ally was the operations manager, who helped me influence the company's CEO about the value of project management. This situation facilitated me spending some time with the company's CEO, having lunch periodically, and sharing with him that he needed to be a sponsor in that project. He listened to me because the Operations manager advised him about; I was a reliable and experienced professional who had implemented similar projects in other organizations successfully and observed my positive attitude.

He told me one day: *Alfonso, you became popular in the company because you smiled early every morning and dealt with everyone with respect.* The CEO believed many organizational processes would improve by starting improvement projects. Then, I had two main allies who were positively contagious at every Board meeting, supporting my action plan. Getting the buy-in from the project sponsor took me a couple of months. Still, finally, he reinforced the importance of the project and encouraged all managers in his organization to follow me as a project leader.

When team leaders were trained in project management, they were assigned as project managers for small projects. At the beginning, project management increased their workload, but in a couple of months, most of them were familiar with the process and documents and were aware of the value. The stress produced at the beginning was diminished little by little with the coaching process I did with all project leaders. We had a meeting every week not only to review the projects progress but also to understand their obstacles and giving them some feedback to improve and move forward. The principle "nothing is impossible" was continuously reminded to all team leaders. Management expected an exponential learning curve, and it was not that way. People need to spend time when they change their work mindset, but they became better and better project management practitioners. I always encouraged their managers to give them positive feedback periodically, creating a culture of sponsorship and support. Managers recognized teamwork and leadership among their teams, and it helped a lot in the project management discipline implementation company wide.

Two years later, I trained team leaders in project management, and upper management supported them. Team leaders managed some improvement projects, and I was their coach, helping them improve and move forward. The initial resistance and issues diminished and became opportunities to enhance the organizational processes. One year later, the company's CEO asked me for a proposal to expand their oil and gas products terminal. It was a ten-million Euro project, and it was successful. So, my optimistic attitude and the application of my passion, persistence, and patience made a difference for this customer converting a couple of years of issues into a business opportunity for both parts.

SUMMARY

- Human beings and professionals appreciate the essential things in life when we live in difficult times.
- The road to success often travels through adversity.
- The project problems and issues you find when managing your projects are always opportunities to learn.
- Ask yourself what you have learned from your trying experience, and focus on moving forward and growing.
- In times of crisis, always strive to maintain an optimistic attitude and an open mind, for this environment will allow you to find the benefit in your difficulty.

TOOL—TURNING PROBLEMS INTO OPPORTUNITIES

Apply the following steps:

- STEP 1: Isolate the specific problem.
- STEP 2: Reframe the situation by getting curious.
- STEP 3: Choose to take full responsibility.
- STEP 4: Ask yourself, "If this problem is a learning opportunity, what is it trying to teach me/us?"
- STEP 5: Decide what you want to replace the problem with.

I found that there are four ways to turn your problems into opportunities:

1. Tap into creativity: Nothing is creative about telling people what to do. But creativity takes passion, persistence, and patience. Creativity requires you to get comfortable NOT knowing. A quick solution is the end of creativity and the beginning of self-justification.
2. Seek elegance: There's nothing elegant about telling people what to do.
3. Develop an opportunity: Anyone can solve a problem. Turning problems into opportunities is leadership.
4. Answer the big question, "What do we want?"

6

Your Words Make a Difference

Repeat anything often enough, and it will start to become you.

–Tom Hopkins

When was the last time you seriously thought about the words you use every day? How carefully do you select them? You might think, "Alfonso, why is there all this fuss about words? What is the big deal?" The answer is simple. Your words often have much more power than you can imagine. They can build a bright future, destroy opportunity, or help maintain the status quo. Your words reinforce your beliefs, and your beliefs create your reality and contribute to project success. Please see Figure 6.1. Think of this process as a row of dominos like this: *Thoughts—Words—Beliefs—Actions—Results*.

Here is how it works. Henry has a *thought*: I am not very good at project sales. He has not thought of this only once. Oh, no. He runs it through his mind regularly, maybe hundreds or thousands of times. Then, Henry starts to use *words* that support this thought. He tells his friends and project management colleagues, "I am never going to do very well in sales," or "I just hate making sales calls or approaching prospects." Here again, Henry repeats these phrases repeatedly in his self-talk and discussions with others. Henry, in turn, strengthens his *beliefs*, and it is at this stage that the rubber meets the road. You see, everything that you will achieve in your life flows from your beliefs. So, in the sales example, Henry believes he will not succeed in project sales, and this thought becomes embedded in his subconscious mind. What can flow from that belief? Because Henry does not believe in his project sales ability, he takes minimal action or actions that are not productive or effective. He does not do the things that would be necessary to succeed in project sales. And then, quite predictably, Henry gets inferior *results*.

 DOI: 10.1201/9781003485674-6

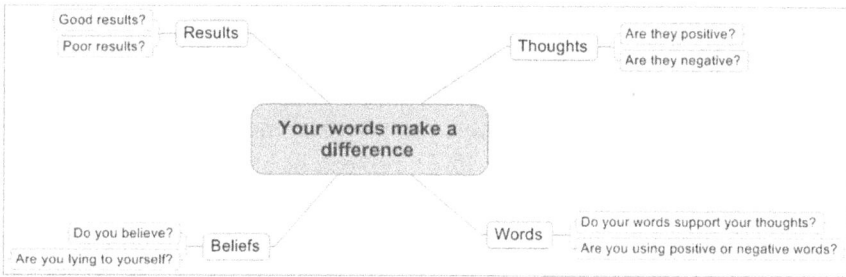

FIGURE 6.1
Your words make a difference.

To make matters worse, Henry then starts to think more negative thoughts, repeat negative words, reinforce negative beliefs, and get even more negative results. It is a vicious cycle. Of course, this whole process could have ended well if Henry had selected positive ***thoughts*** and reinforced them with ***positive words***. In turn, he would strengthen the ***belief*** that he is successful in sales. As a result, Henry would take ***actions*** consistent with that belief and wind up with outstanding ***results***.

SELECTING THE RIGHT WORDS

Do not underestimate the role of your words in this process. Professionals who feed themselves a steady diet of negative words will have a negative attitude. It is a simple matter of cause and effect. You cannot keep repeating negative words and expect to be a high achiever. That is because negative words consistently reinforce negative beliefs, eventually leading to adverse outcomes. We usually repeat to ourselves things like, "I am not good at delivering presentations," "I am not good at talking to upper managers," or "I am not very good at managing project costs." And, after many years of using negative words, you develop a strong belief that you cannot do these things. Do you see how you create this situation by being careless about your words? And the truth is, you could eventually reverse this trend if you start using positive words about your ability to make repairs. This behavior depends on the motivational values of every professional. However, I have observed that many professionals are very pessimistic by nature. If you belong to that group of professionals, now is the moment to wake up and be conscious about the words you use in your projects with your people, customers, and managers. If you are a positive person, congratulations and welcome to the team, "Today is a Great Day!" If you are not yet a member, you can change your attitude. I still remember the first time I met my first supervisor and every virtual meeting I had with him during the first two years of my Ph.D. project; he always gave me suggestions to improve my paper writing and direct feedback, was positive, gave me enough time to reflect, and encouraged me to move forward. I was also so lucky with my second Ph.D. supervisor because she opened every virtual meeting with a big smile; she always provided me with ideas and resources to improve my dissertation writing and encouraged me to move forward.

THE STATE OF MIND

In words, the speaker's state of mind, character, and disposition. Years ago, my friend Randall L. Englund introduced me to Dr. Robert J. Graham, an experienced and worldwide-recognized project professional and a lovely, unique, and friendly person. He is also a cultural anthropologist, a university professor in the US, a consultant worldwide, and the author of five books on project management. Graham has a physical challenge to contend with. Robert Graham is in a wheelchair. He has multiple sclerosis. However, over many years, he has consistently delivered project management seminars, attended and been a lecturer and keynote speaker in many Project Management Congresses, and consulted and assessed big multinational companies in the project management field.

Graham loves people. He always speaks enthusiastically; he transmits passion and security in all speeches. Graham cannot use his arms very much and must rely on a power scooter or a wheelchair. He communicates through positive words, looking at you through his powerful smile. He has told me many stories because he has traveled a lot worldwide. Graham continues traveling coast to coast to visit his family and help with community affairs. Incredibly, someone in his physical state does not focus on himself. He is always trying to enjoy his life, never complaining. Robert Graham empowers himself to achieve great things through positive and enthusiastic words. He does not give any power to his limitations, and as a result, he can transcend them and accomplish more than many other people. What obstacles are you facing in your life right now? Imagine the power you could unleash if you saw them as "just barely an inconvenience" instead of an insurmountable barrier. I have some suggestions to improve your positive words.

To begin with, use positive self-talk as often as possible. In my opinion, the more the merrier. After all, you are talking to yourself, so you don't have to worry about others hearing your comments or judging what you say. The key is that you attend to this positive input repeatedly, and it becomes deeply rooted in your subconscious mind. Whether to share your goals with other people is a much trickier issue. I have learned this: *Never discuss your goals with negative people.* They will only argue and point out why you will not be successful. Who needs that? Often, these "negative nellies" are the ones who do little or nothing in their own lives. They have no goals or dreams, and they do not want anyone else to succeed, either. Yet there are some instances when you can benefit by telling others about your goals. First, speak with someone who is highly positive and supportive of your efforts. It should be the kind of person who would be delighted if you achieved this goal and would do anything in their power to assist you. You may have a friend, colleague, or certain family members who fit this role. Sharing your goals with others working with you is essential to achieve that outcome. For example, if a sales manager wants to increase sales in the coming year by twenty percent, they would make this goal known to everyone on the staff. Then, everyone can work together to achieve it. Even though I am encouraging you to use positive words to move you toward your objectives, I am not suggesting that you ignore the obstacles you may face or discourage feedback from others. Before achieving any goal, prepare for what may be coming. I prefer to discuss those issues with someone cheerful, whose feedback includes creative solutions to possible difficulties.

Furthermore, I will only discuss my plans with qualified people to render an intelligent opinion on the subject. For example, when I left HP to create my own company, I asked some colleagues for *their opinions about my decision. Most of them did not understand. Common opinions were "too risky; you are a fool; outside is raining too much." I only found one professional who encouraged me, Luis M. He supported my ideas and gave me some market contacts to start my business. He was my friend and a reviewer of my first project management book.* I learned to listen to intelligent people focused on the vast possibilities every human being has as a professional. A positive state of mind helped me a lot in my professional career.

WORDS AND ACCOUNTABILITY

There is another reason why, in some cases, you might decide to share your goals with someone else. In other words, if I tell others I am going to do something, I have to go ahead and do it. Think of this approach as "burning your bridges." Let me assure you I am not a believer in "burning bridges" when it comes to personal or business relationships. But sometimes, the only way to move forward in life and to achieve an ambitious goal is to cut off all avenues of retreat. This situation can be a beneficial strategy. We may tell a friend that we will work out at the gym three times this week, knowing that at the end of the week, this friend will ask whether we did go to the gym three times. An even more dramatic example is that of well-known motivational speaker Zig Ziglar. Ziglar decided to go on a diet and reduce his weight from 202 to 165 pounds. At the same time, he was writing his book *See You at the Top*. In the book, Ziglar wrote that his weight was down to 165 pounds. This fact happened ten months before the book went to press. Then, he placed an order with the printer for 25,000 copies. Now, remember, when he wrote these words, Ziglar weighed 202. He put his credibility on the line with 25,000 people. By including a statement that weighed 165 pounds, Ziglar had to lose 37 pounds before printing the book. And he did. Use this strategy selectively. Limit it to those goals that are very important to you and where you are committed to going the distance. Is it risky? You bet it is. But it is a tremendous motivator.

EMOTIONS

Our vocabulary affects our emotions, beliefs, and effectiveness in life. For example, let's say that someone has lied to you. You could react by saying that you are angry or upset. If, however, you used the words "furious," "livid," or "enraged," it would alter your physiology and your subsequent behavior. Your blood pressure would rise. Your face would turn beet red. You would feel tense all over.

On the other hand, what if you characterized the situation as "annoying" or said that you were "peeved"? This situation lowers your emotional intensity considerably. Saying you are "peeved" will probably make you laugh and ultimately break the negative emotional cycle. You would be

Good results?
Bad results? ---- Results

Thoughts ---- Are they positive?
Are they negative?

Your words make a difference

Do you believe?
Are you lying yourself? ---- Beliefs

Words ---- Are your words supporting your ideas?
Are you using positive or negative words?

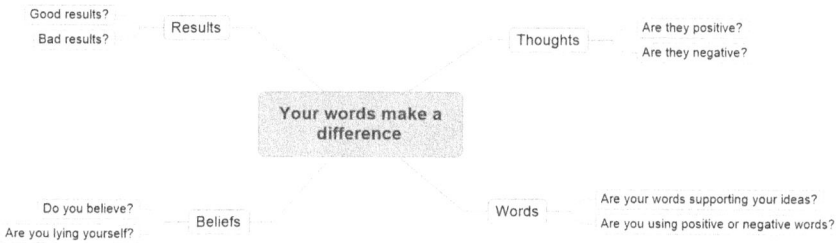

FIGURE 6.2
Your words make a difference.

much more relaxed. Of course, you can also intentionally select words to heighten positive emotions. Instead of saying, "I am determined," why not say, "I am unstoppable"? Or, in place of declaring that you "feel okay," try "I feel phenomenal" or "I feel just tremendous." Juicy, exciting words like that lift your spirits to a higher level and profoundly influence those around you. When you consciously decide to use such terms, you choose to change the path you are traveling. Others will respond to you differently, and you will alter your perception of yourself. Let's take a look at your life for a moment. Are there any areas where you have been using phrases like "I can't," "I am not good at," or "It is impossible?" We all know project professionals who make statements, as shown in Figure 6.2.

When you make these comments daily for ten to twenty years, you are programming your mind for failure. *It all comes back to your attitude.* Every one of these examples reflects a negative attitude. And if you see the world through a smudged window, you will use negative language and get disappointing results. Fortunately, you can control your words, which means you can build a positive belief system and produce the desired results. The first step is awareness. Let's examine the phrases you have used in four key areas of your project life—relationships, finances, career, and health.

1. Relationships
   Do you say things like "All good men (or women) are taking" or "People are always taking advantage of me"? If you do, you are programming yourself for unhappy relationships. Your mind hears every word you speak and sets out to prove you right. About the above examples, your mind will see that you attract only those who will disappoint you or take advantage of you. Is this what you want? If not, *stop repeating such negative statements.*

2. Finances

What words do you use regularly to describe your current financial situation and your prospects for the future? Phrases such as "I am always in debt," "The economy is lousy," or "No one is buying" work against you. Choosing the language that reaffirms prosperity and better economic times is far better. Of course, you will not necessarily have abundant wealth within a few days after changing your language. But the physical conditions can only change after your beliefs have altered. Clearing up your language is an essential first step. After all, the wealthy did not get that way by thinking about poverty. And the people who always talk about a lack of money generally don't accumulate much of it.

3. Career

How would you respond if I asked about your career prospects over the next five to ten years? Be honest. Would you say that things will probably remain the same now? Or would you describe a higher position with more challenges, responsibilities, and increased financial rewards? If you respond, "I don't know where I am going in my career," chances are not much will change. Your language reflects your lack of vision and direction. If, on the other hand, you have a clear goal that you can articulate pretty often, even if only to yourself, you stand an excellent chance of reaching that goal. The same, of course, holds if you have your own business. Do you use language that is consistent with the growth of your business? Or do you constantly talk about how you will never get to the next level?

4. Health

Without question, our words have a profound impact on our health. For example, imagine that a group of us sat down to what seemed to be a perfectly wholesome and delicious meal. Then, two hours later, I called and told you that every person who ate with us had been rushed to the hospital and treated for food poisoning. Suppose that you felt perfectly fine before I called. How would you react after hearing my information? Most likely, you would clutch your stomach, get pale, and feel very ill. Why? My words instilled a belief in you, which, in turn, your body started to act upon. This same bodily reaction would have occurred even if I was playing a cruel joke and was lying about the whole situation. Your body responds to words it hears you and other people say. That's why it makes absolutely no sense to keep repeating that you have "chronic back pain that will never go away" or that you get "three or four bad colds every year." By uttering these statements, you instruct your body to manifest pain and disease. Please don't misunderstand. I am not suggesting that you deny pain or disease or that you can overcome any illness. You do not gain anything using language that reinforces suffering and incurability.

UP TO YOU

Have you considered the words you use in your personal and professional life? When we repeat specific phrases repeatedly, it is as if a "groove" is formed in our brain. We keep replaying the same old refrain like a broken record. Whenever you say these words, the trouble is you deepen the groove, replaying the same ancient myths in your mind, strengthening the same old beliefs, and getting the same old results. Recognize, however, that there is no reason to unquestioningly continue doing so just because you have said things in the past. While it takes some discipline and vigilance to change your language, it is well worth the effort. So, from now on, consciously choose words that will point you toward your goals. Ask a friend to remind you when you slip. Remember, it is up to you to speak in a way that will move you toward what you want in life and the projects you manage.

NONVERBAL COMMUNICATION

Nonverbal messages often contradict the verbal; often, they express feelings more accurately than spoken or written language. Numerous articles and books exist on the importance of nonverbal messages. Some studies suggest that sixty to ninety percent of a message's effect comes from nonverbal cues. I mean:

- Appearance
- Body language
- Silence, time, and space

Appearance: It conveys nonverbal impressions that affect receivers' attitudes toward the verbal message before they read or hear it. For instance, an envelope's appearance—size, color, weight—may impress the receiver as "important," "routine," or "junk" mail. Next, the letter, report, or title page communicates nonverbally before reading its contents by the type of paper used, length, format, and neatness. Finally, the language itself, aside from its content, communicates something about the sender. Is it carefully worded and generally correct in mechanics such as spelling, grammar, and punctuation?

What is the effect on oral messages? Your appearance and surroundings convey nonverbal stimuli that affect attitudes toward your spoken words, whether speaking to one person, face to face, or a group in a meeting. Personal appearance: Clothing, hairstyles, neatness, jewelry, cosmetics, posture, and stature are part of personal appearance. Depending on circumstances, they convey impressions regarding occupation, age, nationality, social and economic level, job status, and excellent or poor judgment. Appearance of surroundings: Surroundings include room size, location, furnishings, machines, architecture, wall decorations, floor, lighting, windows, and view. Surroundings will vary according to status, country, and culture. Body language: Included under body language are facial expressions, gestures, posture and movement, smell and touch, and voice and sounds.

The eyes and face are beneficial means of communicating nonverbally. They can reveal hidden emotions, including anger, confusion, enthusiasm, fear, joy, surprise, and uncertainty. They can also contradict verbal statements. For example, because he was embarrassed, a new team member answered "yes" when asked by his project manager if he understood his instructions. Yet the project manager should have noticed the employee's bewildered expression and hesitant voice and restated the instructions more clearly. In the US, direct eye contact, but not staring, is desirable when two people talk. The person whose eyes droop or shift away from the listener is considered shy, dishonest, and untrustworthy. However, we must remember that it differs in other cultures and depends on the situation. In some occupations, actions speak louder than words. Gestures and movements are culture-specific. The meaning of a gesture in the US may be completely different in Europe and Asia. A clenched fist pounding on the table in the US can indicate anger or emphasis. Such a display in Asia would be unacceptable. Continual gestures and movements may signal nervousness and distract listeners. Handshakes reveal attitudes, sometimes handicaps, by their firmness or limpness.

Legs, too, communicate nonverbal messages. Consider, for example, a man sitting with his legs stretched out on his desk during an interview, a person shifting from one leg to the other in rhythmic motion, or a woman pacing back and forth while speaking. Posture and movement can convey self-confidence, status, or interest. A confident executive may have a relaxed posture and stand more erect than a timid subordinate. An interested listener may lean forward toward the speaker; one who is bored may lean away, slump, or glance at the clock. Various odors and fragrances sometimes convey the sender's emotions and affect the receiver's reactions, especially if the receiver is sensitive to scents. Also, touching people can communicate friendship, love, approval, hatred, anger, or other feelings. Your voice quality and the extra sounds you make while speaking is also a part of nonverbal communication called paralanguage. Paralanguage includes voice volume, rate, articulation, pitch, and the other sounds you may make, such as throat clearing and sighing. The words "You did a great job on this project" could be a compliment. But if the tone of voice is sarcastic and said in the context of criticism, the true meaning is anger. A loud voice often communicates urgency, while a soft one is sometimes calming. Speaking fast may suggest nervousness or haste. A lazy articulation, slurring sounds, or skipping over syllables or words may reduce credibility. A lack of pitch variation becomes a monotone, while too much variation can sound artificial or overly dramatic. Throat clearing can distract from the spoken words.

Emphasizing certain words in a sentence can purposely indicate your feelings about what is essential. Silence, time, and space can communicate more than you may think, even causing hard feelings and loss of business and profits. It pays to know these differences across cultures. Suppose you wrote a request to your supervisor for additional funds for a project you are developing. If you receive no answer for several weeks, what is your reaction? Do you assume that the answer is negative? Do you wonder if your supervisor is merely very busy and has not been able to answer your request? Do you think your supervisor is rude or considers your request unworthy of an answer?

Concepts of time, however, vary across cultures and even in the US. Americans and Germans, for example, are pretty punctual. Middle

Eastern business people think little of arriving after an agreed-upon time, not out of discourtesy but rather a feeling that they will accomplish the task regardless of time. If you arrived on time for a meeting in Spain, your host might wonder why you came so early.

CASE STUDY 1

Some years ago, I managed a project for the Spanish Foreign Ministry. It was an infrastructure project focused on creating the proper hardware and software architecture to encode and decode messages between worldwide Embassies. The IT manager (José) had obvious requirements for that project but never wrote them down. We did it for him. I thought he was problematic, vulgar, and prone to speaking and acting out negative behavior. I still remember that he said I was always smiling, and he was disappointed. However, his people worked well with me and my team. We took care of using respectful sentences working with them, and they appreciated the difference between their boss and us. One day, we had a database problem and needed to work overnight with the whole team. The customer IT manager was not there, but when he returned the following day, his first sentence was: "Hello, are you working or talking as always?" Can you imagine the feeling of everyone who heard that sentence?

The IT customer was happy about the results, and we had lunch together. At the end of the lunch, the IT manager said to the team: "Congratulations on the project's success, but you delivered the project over." I continued working for that customer with my team on other projects. After some years, I gained good credibility with this customer, and occasionally, we went for lunch together. In one of the lunches, he told me, "Alfonso, my people respect you very much. They are happy when you come to my office. What do you do?" I answered him, "I am listening to them, respecting them, and telling them that I need them to achieve project success. I smile frequently because every day is a great day." I told him, "José, spend more time with your people, talk to them, don't be negative with them, be more aware of the words you use, and appreciate their efforts and achievements personally. The return on your investment will be significant." I feel happy about this story because now this customer meets me with a big smile whenever I see him. He has changed his behavior to be more positive, and his people greatly appreciate him for it.

CASE STUDY 2

Since 1993, I had the opportunity to share my project management experiences at different conferences and workshops. I am very enthusiastic and speak with high energy when speaking in front of the public. I used to share my professional and personal stories with the audience. I believe the words I use make a difference because I always speak from my heart, being authentic when dealing with people. More than once, after delivering one of my speeches, some people approached me and told me that I inspired them. I know I am not a magician but a project management practitioner who cares about people. Pronouncing the right words at the right moment is precious; sometimes, when the project manager is stressed, they cannot control their words. Talking badly to free adrenaline is easy, but controlling your emotions and being worried about how they may affect the people around you is complicated, but it is not impossible.

I have often felt stressed in my professional career managing projects for different reasons. Let me share the following story: I worked as a project manager for an IT customer project far from home. The project lasted almost two years, and in the middle of it, I had a lot of pressure from the customer site and my boss. After a work meeting at the customer site, I lost my nervousness and threw myself to the floor. People in the room were shocked by my reaction. Immediately, I went to the restroom, washed my face, drank water, and returned to the meeting room. Then, I apologized to everyone in the room. I was out of control, but I asked people to forgive me, "I am a simple human being; my sincere apologies; I could not put up with that pressure. Since then, people have known me better and helped me move forward; my words have made a difference in all of them.

We are often unaware of people's perceptions about some words and make many mistakes. My project management passion always helps me search for the appropriate words when I feel any project stakeholders frustrated, disappointed, or worried. I am not always successful in my search, but at least I use my persistence.

SUMMARY

- Remember that your words often have much more power than you can imagine. They can build a bright future, destroy opportunity, or help maintain the status quo. Your words reinforce your beliefs, and your beliefs create your reality and then contribute to your project.
- You cannot keep repeating negative words and expect to be a high achiever. Negative words consistently reinforce negative beliefs, eventually leading to adverse outcomes.
- Our vocabulary affects our emotions, beliefs, and effectiveness in life. Fortunately, you can control your words, which means you can build a positive belief system and produce the desired results. The first step is awareness.
- Remember, it is up to you to speak in a way that will move you toward what you want in life and the projects you manage.

TOOL—ASSESS THE WORDS YOU ARE USING

Please respond to the following questions and follow the provided suggestions:

1. Do you have positive or negative thoughts daily?
2. Do you have more positive than negative thoughts?
3. Are you using positive words with your peers, project stakeholders, and others daily? Please list them at the end of the day.
4. Are you using negative words with your peers, project stakeholders, and others daily? Please list them at the end of the day.
5. Please write down your positive thoughts and read them at the end of the day.
6. Please write down your positive words and read them at the end of the day.
7. REPEAT this process for twenty days.

7

How Are You?

Your day goes the way the corners of your mouth turn.

–Unknown

Our answer to the question: How are you? It seems like such a small thing. But we must answer that question many times every day. Then, it is not a small thing at all. It is a significant part of our daily conversation. When someone asks: How are you? What do you say? Your answer is usually no more than a few words. And yet, that short response tells a lot about you and your attitude. Your response can shape your attitude. I have observed that the "How are you?" responses may be harmful, mediocre, and positive. Let's examine these three categories and some common responses under each one.

NEGATIVE ANSWERS

The negative reply to "How are you?" may include phrases, as you can see in Figure 7.1.

When a project manager professional answers with *"Don't ask,"* I know I am in for trouble. That person will unleash a multitude of complaints and make me sorry for asking the question in the first place. And I pity those who take the "Thank God it is Friday" approach to life. They say "Monday, Tuesday, Wednesday, and Thursday" are bad weekly days. For these people, four-fifths of their work week is lousy. The fifth day, Friday is "bearable" only because they know they will have the next two days off. Is this a way to live your projects and your life? Are you beginning to see how these negative phrases can poison your attitude and turn off your project team and other project stakeholders?

DOI: 10.1201/9781003485674-7

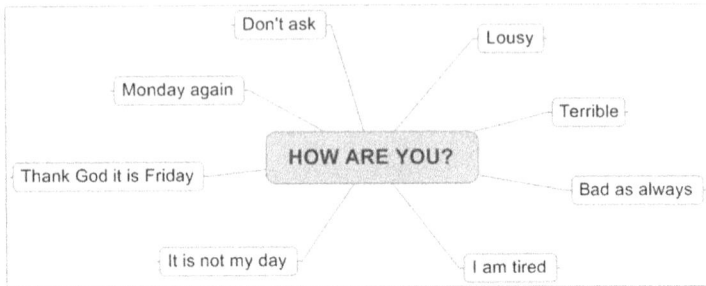

FIGURE 7.1
How are you?

POOR ANSWERS

Those in the mediocre group are a step up from the negative bunch but still have room for improvement. Here are some of the things they say:

- "I am OK."
- "Not too bad."
- "It could be worse."
- "Every day older and older."
- "I am fine."

Do you want to spend much time with someone who thinks life is "not too bad"? Is that the person you like to do business with? When we use words like these, we also diminish our energy. Can you imagine someone saying "could be worse" with an upright posture and a lot of enthusiasm? Of course not. These people sound like they have not slept in two days. There is no getting around it. People who use mediocre words will develop a modest attitude and get mediocre project results. And I know you don't want that.

POSITIVE ANSWERS

Passionate people give enthusiastic responses, like:

- "I am Terrific."
- "I am Fantastic."

- "Today is a great day."
- "Great."
- "I am Excellent."
- "Superb."
- "If I were better, I would have a twin."

Those who use positive words like these have a bounce in their step. You feel a little better just by being around them. Be honest. How did you feel as you read the positive list? I don't know about you, but I am energized and excited as I review that list. These are the people I look forward to meeting today. These are the people who are more likely to get my business. Why not go back and re-read the negative list and the mediocre list? Say them out loud. How do they make you feel? You see, if given the choice, I would rather be around people who are cheerful and full of life than those who are negative and listless. It is like the old saying that everybody lights up a room—some when they walk into it and some when they walk out. You want to be the one who lights up a room when you walk in. As for me, when someone asks me how are you? I usually respond, "Very good, today is a great day." It projects a positive attitude to the other person; the more I say it, the better I feel.

JOIN THE POSITIVE PROFESSIONALS

Well, you have had a chance to review some typical responses in each category—negative, mediocre, and positive. Which of these phrases do you use most often? Which responses do your friends and family use? If you find yourself in the negative or mediocre group, I suggest you immediately consider revising your response and joining the ranks of the positive. Here is why. When somebody asks how you are? And your physiology is adversely affected if you say horrible or not too bad. You tend to slump your shoulders and head and take on a depressed posture. What about your emotions? After stating that you are lousy, do you feel better? Of course not. You feel even more down in the dumps because negative words and thoughts generate negative feelings and, eventually, negative results. It is up to you to break it. Even if actual circumstances in your life persuaded you to state that you are lousy—perhaps a promising business deal fell through, or your child received poor grades in school—your gloomy attitude does nothing to improve the situation. If you want to make matters worse, your mediocre or negative reply turns others off; they are dragged down just being around you and hearing your pessimism.

PRACTICE A NEW APPROACH

If all of these negative consequences flow from your words, why do you continue to say them? More than likely, it is because you have not recognized that you have a choice in the matter. Instead, you are following a habit you developed many years ago that no longer serves you. In the end, your own words are a self-fulfilling prophecy. If you say, "Everything is terrible," you are attracting those people and circumstances that will cause that statement to be true. If, on the other hand, you repeatedly state that your life is lovely, your mind will begin to move you in a positive direction. For instance, consider what happens when you respond that you are "Excellent or Terrific." As you say these words, your physiology corresponds with your optimistic language. Your posture is more upright. Other people are attracted to your energy and vitality. Your business and personal relationships improve. Will all of life's problems magically disappear? No, but you have set a fundamental principle: We get what we expect in life.

I can tell you from firsthand experience that this is one of those little things in life that makes a big difference. About twenty-five years ago, when someone asked me how are you? I said something like, "OK," with very little energy. What was I doing? I was programming myself to have "okay" relationships with people. I was programming myself to have "okay" success. I was programming myself to have an "okay" attitude and "okay" life. But then, through an experiment, I learned that I didn't have to settle for an "okay" life. I picked up my response a few notches and said, "Terrific." I told it with energy. Sure, at first, it was a little uncomfortable. Some people looked at me like I was a little strange. But after about a week, it started to come naturally. I was amazed at how much better I felt and how people were more interested in talking with me. I believe that this is not rocket science. You don't need talent, money, or good looks to have a great attitude. It would help if you got in the habit of using a high-energy, positive response, and you will reach the same exciting results I got.

WHAT HAPPENS IF I FEEL THAT TODAY IS NOT A GREAT DAY?

Whenever I do a presentation or a Seminar, I say, "Today is a great day." Many people are surprised at the beginning. Sometimes, at the end of the event, some people came to me and asked, "What if today is not a great day?" I don't want to lie to my customers and colleagues by telling them everything is lovely when it is not. Don't get me wrong: I put the highest value on integrity and telling the truth. Yet I don't think this is a matter of speaking the truth. Let me explain. Assume for a moment that Rose feels tired. When someone at work asks her how she is? She wants to be perfectly honest and says, I am exhausted. Here is what will happen.

Rose will reinforce the belief that she is exhausted. She will feel even more fatigued. Rose will probably slump her shoulders and let out a sigh. She will have a lousy, unproductive day at work. Let's get back to the person who asked Rose the question—and who probably regrets it now. That person also feels worse. After all, when someone tells you how tired she is, do you feel uplifted? No way. Just suggest the world being "tired," and you start yawning. So, Rose has brought herself down, as well as her co-worker. OK, Rose goes home after her grueling day and is now exhausted. So, she plops into her favorite chair and opens the newspaper to look at the winning lottery numbers. Rose has the winning ticket. She just won $10 million.

What do you think Rose would do? Remember, she is exhausted. You and I know that Rose would leap out of her chair, jumping up and down, screaming, and waving her arms. You would think she was leading an aerobics class. Rose would run to pick up the phone to call her family and friends. She would be a bundle of energy and would probably stay up all night celebrating and planning what to do with the money. But wait a second. Ten seconds ago, Rose was exhausted. Now, she has the energy of a fifteen-year-old cheerleader who somebody told she made the cheerleading squad. What happened in those ten seconds to change someone from utterly exhausted to wildly exuberant? Did she get a shot of vitamin B-12? Did anyone throw a bucket of ice water in her face?

No. *Rose's transformation was entirely mental!*

Now, I am not trying to diminish what Rose was feeling. Her fatigue was genuine but not as much physical as mental. So, was Rose telling the truth when she said she was tired? It has very little to do with the truth. It is a matter of what Rose wants to focus on. She could concentrate on feeling tired. That was one option.

On the other hand, she could have thought about the many blessings in her life and felt very fortunate and energized. How we think is very often a subjective matter. When we tell ourselves that we are tired, we feel exhausted. When we tell ourselves that "today is a great day," we feel energized. We become what we think about.

USE YOUR PASSION TO ANSWER

Try this experiment. When anyone asks How are you? Respond with energy and enthusiasm that you are Great or Today is a great day! Say it with a smile and a sparkle in your eye. It does not matter whether you feel terrific at that moment. Apply the act-as-if principle. In other words, if you want to be more positive, act as if you already are, and pretty soon, you will find that you have become more positive. Do not worry if you are uncomfortable saying these words at the beginning. Stick with it, and eventually, you will grow into it. You will quickly notice that you feel better, that others want to be around you, and that positive results will come your way. Then, "How are you?" Today is a great day.

CASE STUDY 1

At thirty-three, I was a project manager working on a project far from my city. When some manager asked me about my project, I always answered: "Oh my God! I am so frustrated because of the lack of support from my organization for this particular project. I also feel frustrated with my customer; he is not collaborating." All those answers were negative, but I was not conscious of that. I was unaware that my team was also listening to my answers. They observed my negative attitude. I generated a lot of frustration and lack of motivation in my team, and I didn't know it. I was fortunate. One weekend, I met a manager from another organization in a restaurant and had the opportunity to drink a cup of coffee with him. I told him about my project situation. He gave me great ideas and advice to move forward and improve my attitude. He said: "*Alfonso, there is no project without issues or problems. It is like life; the point is that you need to deal with those problems and not forget that you work for an organization; you must ask them for their support. You must communicate with all project stakeholders and tell them how much you need them to achieve project success. I am sure you will have low moral moments, but please be focused on the great things you have. Think about your family, which is OK. Think you can do it, and be positive. When somebody asks you about your project, say: 'It is under control. I have some issues to manage, but if I need help, I'll let you know'. That answer is more optimistic; it inspires optimism, energy, and enthusiasm. Many times, there is a lack of power in some projects.*"

That answer empowered me a lot. It was not easy to use that advice immediately, but I did it step by step. I apply my approach every day because "Every day is a Great Day!" You will surely be able to choose between the positive part of your project or the negative one. I suggest you select the "positive." After more than forty years of practical experience, every day is good for me because I always try to focus on the whole part of the bottle. Please practice my philosophy, and you will be happier.

CASE STUDY 2

Since January 2023, I have been dealing with an incurable disease. Many people think that it is not worth to make an effort and maintain a positive spirit, but I think that it is not the end, by now. Some colleagues think I would stop my project management activities because I have cancer, and I do not think so. I did not because I need to be thankful, I am still alive. I recharge my batteries when doing project management activities every day, I feel useful, and my pains seem less. I am delivering seminars, speeches, virtual workshops, writing a new book, could finish my Ph.D., and I am teaching some training courses at universities, obviously not every day, neither every week, but I maintain my brain and body active. I need that.

When people are asking me, "How are you?" I always answer I am fine. People answer me, "Do you have any pains?" and I answer yes, I have some during the day, but they are part of my life now. It is part of my medical treatment, I am aware of my situation, so "TODAY IS A GREAT DAY, and TOMORROW WILL BE EVEN BETTER."

I try to enjoy every day, and transmit positivism to all around me. Some people cannot understand my positive daily reactions and why I am not constantly complaining about my pains. Some days, I would not get up because I am tired from early in the morning, but I get up and say thank you for one more day alive. I need to be grateful every day in life; I can still contribute to project managers' happiness by conveying the message that nothing is impossible except passing away.

I had the opportunity to teach project management soft skills to a group of thirty university students. It was a special effort because I delivered in the evening, and I was tired, but at the moment I started my class, everything changed in my mind, I felt stronger and happy. Only seeing the engaging faces of my students when I shared my experiences and stories was amazingly good. What a nice atmosphere in the classroom when people are participating and living an amazing experience. I also learned a lot from them, from their creativity, ideas, and experiences too.

How are you? I am terrific, excited, and happy when I can care of people's behaviors, and I am able to create a positive environment of collaboration and learning. It is also the reason because I am writing this book, because everything is possible in life if you believe that it is. Make people happy, and you will be happy too.

SUMMARY

- When someone asks: How are you? What do you say? Your answer is usually no more than a few words. And yet, that short response tells a lot about you and your attitude. Your response can shape your attitude.
- As for me, when someone asks me how are you? I usually respond, "Very good, today is a great day." It projects a positive attitude to the other person; the more I say it, the better I feel.

- If you find yourself in the negative or mediocre group, I suggest you immediately consider revising your response and joining the ranks of the positive.
- You don't need talent, money, or good looks to have a great attitude. It would help if you got in the habit of using a high-energy, positive response, and you will get exciting results.
- How we feel is very often a subjective matter. When we tell ourselves that we are tired, we feel exhausted. When we tell ourselves that we feel terrific, we feel energized. We become what we think about.
- Do not worry if you are uncomfortable saying these words at the beginning. Stick with it, and eventually, you will grow into it. You will quickly notice that you feel better, that others want to be around you, and that positive results will come your way. Then, "How are you?" Today is a great day!

TOOL—HOW ARE YOU—ASSESSMENT

The following tool will help you evaluate "how do you feel" now. Please select one of each group, analyze your answers, and you will know which group you are in, and prepare a plan to improve.

Negative

- Lousy.
- Terrible.
- Bad as always.
- Tired.
- It's not my day.
- It's Friday.
- Monday again.
- Could you not ask me?

Poor

- OK.
- Not too bad.

- It could be worse.
- Older and older.
- Fine.

Positive

- Terrific.
- Fantastic.
- Today is a Great Day!
- Great.
- Excellent.
- Super.
- If I were better, I'd have a twin.

8

Don't Complain About Your Projects

If you have time to whine and complain about something, then you have the time to do something about it.

–Anthony J. D'Angelo

My father said that a pessimistic person is an optimistic one who is well-informed. How do you feel when someone unloads all his problems and complaints to you? Not very happy and energized. The truth is, nobody likes to be around a complainer—except, perhaps, other complainers. Of course, all of us complain at one time or another. The critical question is: How often do you complain? If you wonder whether you complain too much, ask your colleagues. They will let you know. Now, when I say "complain," I am not talking about those instances when you discuss your problems in an attempt to search for solutions. That's constructive and commendable. I am not referring to those occasions when you share your project life experiences with colleagues or friends in the context of bringing them up to date on the latest developments in your professional life. After all, part of being human is sharing our experiences and supporting each other.

NOBODY WANTS TO HEAR ABOUT YOUR PAINS

One of the most common areas of complaint is the subject of illness. In this category are comments such as "My back is killing me" or "I have a terrible headache." Worse yet, some people get very graphic in explaining the gory details of their particular ailment. What can I do for you if you have a stomach ache? I am not a physician; you must go to the doctor if you have

DOI: 10.1201/9781003485674-8

a medical problem. More importantly, why are you telling me this? You might want sympathy, but all you are doing is dragging me down and reinforcing your suffering. Talking about pain and discomfort will only bring you more of the same and encourage those around you to look for the exits.

Regarding complaints about projects, the principle of escalation usually rears its ugly head. Here is how it works. You tell your colleague about the problems you have in your project. Your friend interrupts and says, "You think you had it bad. When I managed my last project, the first day, I had a 40 Celsius degree fever and had to go to the hospital. I almost died." Or, tell someone that your back or foot hurts—and count how many seconds it takes for that person to switch the conversation to their back pain and aching feet. Complainers love to play this game; their pain is always worse than yours.

SOME PEOPLE HAD SOME REASONS TO COMPLAIN

I was in my office and thinking about some of the activities of my project that were not going as well as planned. You know the typical project problems—results not happening as fast as expected. And I'll confess that I had been doing a little complaining about it. Then José walked in. José is in his early thirties and came to Spain from Nicaragua about six years ago. He works for a company that cleans homes and offices. You talk about a positive attitude; José is one of the most positive people I have ever met. He is always smiling and upbeat. On this day, however, I asked José about the last earthquake and its impact on his homeland. The smile quickly left his face. He told me of the devastation the earthquake had caused. Thousands of people had died. José said that his father, mother, and brother still lived in Nicaragua, and he had no idea if they were dead or alive. He had no way to contact them. The earthquake destroyed all phone lines in that area. José said he thought about his family every day. Can you imagine the agony of not even knowing if your family is still alive?

Then José went on to tell me about all the things he was doing to help the people in Nicaragua. He was collecting money, clothing, and other necessities. He was actively working with the relief organizations. Instead of just gripping about the problem, he was doing whatever he could do to ease their pain. After speaking with José, I realized how inconsequential my problems were and how fortunate I am. You better believe I stopped complaining. I faced the rest of the day with renewed energy and a better attitude. By the way, several weeks later, I saw José again. And yes, he had his usual winning smile and his great attitude. The good news is that his family members are all alive. The bad news is that they lost everything in the earthquake. I cannot even fathom what it is like to lose everything you own and start over from scratch, especially under these challenging conditions. José has every reason to complain about his family's bad luck. But he does not. He realizes complaining would be a terrible waste of time and energy. Thank you, José, for reminding us that complaining is not the answer to our challenges in life.

PUTTING THINGS IN PERSPECTIVE

There is another valuable lesson that we can learn from José, and that is the importance of keeping things in perspective. Over the years, I have noticed that complainers lack perspective; they tend to blow their problems out proportionately. Optimistic people tend to understand what is truly important in life. Think about the people you know. Do you have friends who get bent out of shape because they have a flat tire?

These folks have lost sight of the "relative importance" of things. We can all learn from Eddie Rickenbacker, who drifted in a life raft for twenty-one days, hopelessly lost in the Pacific. After surviving the ordeal, Rickenbacker said, "If you have all the fresh water you want to drink and all the food you want to eat, you ought never to complain about anything." Let me share with you some of the things I am grateful for. See Figure 8.1.

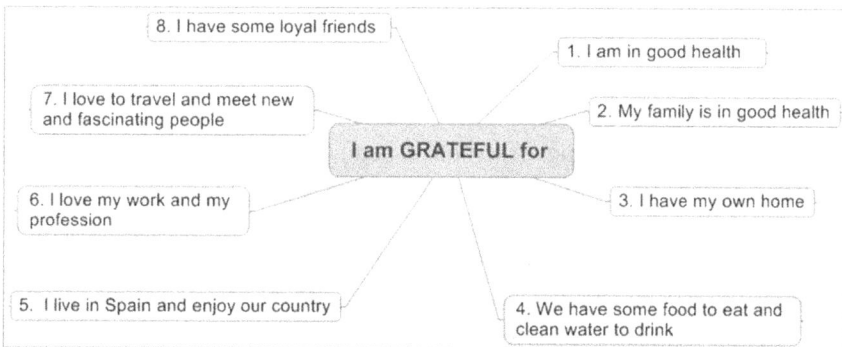

8. I have some loyal friends

1. I am in good health

7. I love to travel and meet new and fascinating people

2. My family is in good health

I am GRATEFUL for

6. I love my work and my profession

3. I have my own home

5. I live in Spain and enjoy our country

4. We have some food to eat and clean water to drink

FIGURE 8.1
I am grateful for.

This mind map shows a partial list of the blessings in my life. Even with all of these beautiful things, there are times when I start to take some of them for granted. But I have learned to quickly re-connect with these blessings, which boost my attitude and bring me back on course. So, what is it that you have been complaining about lately? Are they really "life and death" matters? The next time you feel tempted to gripe about your problems, pick up a pen and piece of paper and start listing all the reasons you must be grateful. Then, put all your project issues and problems in perspective.

THE POSITIVE NEWS GENERATOR

I am not suggesting that you sit back and ignore all of the problems in your life. However, rather than complaining, it is far better to focus your attention and energy on those steps you can take to solve or at least lessen your problem. For instance, let's say you are feeling a little tired lately. Instead of telling everyone how lousy you feel, make an effort to exercise more regularly or get to bed a little earlier. To review: Complaints work against you in three ways. First, no one wants to hear negative news about your illness and your problems. Second, complaining reinforces your pain and discomfort. So why keep replaying painful, negative memories? Third, complaining accomplishes nothing and diverts you from the constructive actions you could take to improve your situation.

I observe that 90 percent of the people don't care about your problems, and the other 10 percent are glad you have them. Seriously though, all of us can cut down on our complaining. From now on, let's do ourselves and others a favor and make our conversations uplifting. The people who don't complain very much (and those who speak positively) are a joy to be around. Decide to join that group—so people won't have to cross the street when they see you coming. The most crucial step in quitting complaining is disconnecting the undesirable behavior from your identity. A common mistake chronic complainers make is to self-identify with the negative thoughts running through their minds. Such persons might admit, "I know I'm responsible for my thoughts, but I don't know how to stop myself from thinking negatively so often." That seems like a step in the right direction, and to a certain degree, it is, but it's also a trap. It's good to take responsibility for your thoughts, but you don't want to identify with those thoughts to the point you end up blaming yourself and feeling even worse. A better statement might be, "I recognize these negative thoughts going through my mind. But those thoughts are not mine. As I raise my awareness, I can replace those thoughts with positive alternatives." You have the power to recondition your thoughts, and the trick is to keep your consciousness out of the dilemma of blame. Realize that while these thoughts flow through your mind, they are not you. You are the conscious conduit through which they flow.

MENTAL CONDITIONING

Although your thoughts are not you, if you repeat the same thoughts repeatedly, they will condition your mind to a large extent. It's almost accurate to say that we become our dominant thoughts, but that's taking it too far. Consider how the foods you eat condition your body. You aren't going to become the next meal you eat, but that meal will influence your physiology, and if you keep eating the same meals over and over, they'll significantly impact your body over time. Your body will crave and expect those same foods. However, your body remains separate and distinct from the foods you eat, and you're still free to change what you consume, which will gradually recondition your physiology by the new inputs. If you keep holding negative thoughts, you condition your mind to expect and even crave those continued inputs. Your neurons will even learn to predict the reoccurrence of negative stimuli. You'll practically become a negativity magnet.

THE TRAP OF NEGATIVE THINKING

Negative thinking is a challenging situation to escape because it's self-perpetuating, as anyone stuck in negative thinking knows all too well. Your negative experiences feed your negative expectations, attracting new negative experiences. In truth, most people who enter this pattern never escape it in their entire lives. It's just that difficult to escape. Even as they rail against their negativity, they unknowingly perpetuate it by continuing to identify with it. If you beat yourself up for being too negative, you're simply reinforcing the pattern, not breaking out. Most people in this trap will remain stuck until they experience an elevation in their consciousness. They have to recognize that they're trapped and that continuing to fight their negativity while still identifying with it is a battle that will never win. Think about it. If beating yourself up for being too whiny was going to work, wouldn't it have worked a long time ago? Are you any closer to a solution for all the effort you've invested in this plan of attack?

Consequently, the solution I like best is to stop fighting and surrender. Instead of resisting the negativity head-on, acknowledge and accept its presence; this will raise your consciousness. Post the acronym WIO everywhere in clear sight to remind you that "Whining Is Optional."

OVERCOMING NEGATIVITY

You can learn to embrace the negative thoughts running through your head and transcend them. Allow them to be, but don't identify with them because those thoughts are not you. Begin to interact with them like an observer. I would say that the mind is like a hyperactive monkey. The more you fight with the monkey, the more hyper it becomes. So, instead, relax and observe the monkey until it wears itself out. Recognize also that this is why you're here, living out your current life as a human being. Your reason for being here is to develop your consciousness. If you're negative all the time, your job is to expand your consciousness to the point where you can learn to stay focused on what you want and to create positively instead of destructively. It may take you more than a lifetime to accomplish that, and that's okay. Your life is always reflecting to you the contents of your consciousness. If you don't like what you're experiencing, your skill at conscious creation remains underdeveloped. That's not a problem because you're here to develop it. You're experiencing what you're supposed to be experiencing, so you can learn.

CONSCIOUS CREATION

You're free to take your time to work through your negativity if you need a few more lifetimes. Conscious creation is a big responsibility, and you may not feel ready for it yet. So, until then, you will perpetuate the pattern of negative thinking to keep yourself away from that realization. You must admit that being the primary creator of everything in your current reality is a bit daunting. What are you going to make of your life? What if you screw up? What if you create a big mess of everything? What if you try your best and fail? Those self-doubts will keep you in a pattern of negativity to avoid that responsibility. Unfortunately, this escapism has consequences. The only way actual creators can deny responsibility for their creations is to buy into the illusion that they aren't creating them. This way means you have to turn your creative energy against yourself. You're like a god using these powers to become powerless. You use your strength to make yourself weak. You may be stuck in a negative thought pattern because you chose it at some point. You figured the alternative of accepting full responsibility for everything in your reality would be worse. It's too much to handle. So, you turned your thoughts against yourself to avoid that tremendous responsibility. And you'll remain in a negative manifestation pattern until you're ready to accept some of that responsibility back onto your plate. Negativity needn't be a permanent condition. You still have the freedom to choose otherwise. In practice, this realization typically happens in layers of unfolding awareness. You begin to accept and embrace more and more responsibility for your life.

ASSUMING TOTAL RESPONSIBILITY

You see, the real solution to complaining is responsibility. You must say to the universe (and mean it), "I want to accept more responsibility for everything in my experience." Here are examples of what I mean by accepting responsibility:

- If I'm unhappy, it's because I'm creating it.
- If a problem bothers me, I'm responsible for fixing it.
- If someone is in need, I'm responsible for helping them.
- If I want something, it's up to me to achieve it.
- If I want certain people in my life, I must attract and invite them to be with me.
- If I don't like my present circumstances, I must end them.

On the flip side, it may also help to take responsibility for all the good in your life. The good stuff didn't just happen to you. You created it. Pat yourself on the back for what you like, but don't feel you must pretend to enjoy what you don't want. But do accept responsibility for all of it to the extent you're ready to do so. Complaining is the denial of responsibility. Blame is just another way of excusing yourself from being responsible. But this denial still wields its creative power. Conscious creation is indeed a tremendous responsibility. But in my opinion, it's the best part of being human. There's no substitute for creating a life of joy, even if it requires taking responsibility for all the unwanted junk you've manifested up to this point. When you complain, stop and ask yourself if you want to continue denying responsibility for your reality or to allow a bit more responsibility back onto your plate. Maybe you're ready to assume more responsibility, and perhaps you aren't, but do your best to make that decision consciously. Do you want sympathy for creating what you don't like or congratulations for creating what you do want?

CASE STUDY

The first thing that I do every day is to thank God for one more day of life in earth. Every project manager and every human being may have some reasons to complain about different situations, for example, when they do not have management support, or they are not understood by their customer, or even worse if they have an illness, or lose a relative or friend because they feel like living an unfair situation in life, we need to accept our reality, and in many occasions, we are unconscious about it. When you complain, stop and ask yourself if you want to continue denying responsibility for your reality or allow a bit more responsibility back onto your plate. We cannot change the course of many things, but we can face them in a positive way.

Some years ago, the organization I worked for assigned me as a project manager for an important IT project far away from home. The project lasted one and a half years, and I had to travel 400 kilometers every weekend and stay there for the whole week during many months. As I had a family of three children and my wife, it was a challenge for me to stay many months far away. On the other hand, it was a professional opportunity for me as a project manager; my organization was confident on me to manage that important project. At the beginning, I complained to my colleagues because I thought it was an unfair situation. My youngest son was only 10 months old, and it would be difficult for my wife to take care of all three every week for several months. But I never complained to my management team and I accepted my responsibility with courage. It was a challenging project which consisted on implementing a new open system platform, customizing a software package from a third party, and managing the organizational change produced by that project in the customer organization. One organizational consulting company (third party) was hired to be part of my team, but it was imposed by the customer IT manager, so I needed to be flexible dealing with their modus operandi and accepting that constraint. So I had another reason to complain, but instead of that I dealt with the team members of that consulting company in a positive way. As organizational consultants they had the trend to join only consultants and not technical people at lunch or coffee breaks, but I broke that trend by treating his boss for breakfast frequently; two months later I felt that they felt in the same boat than my direct team. I could say that after that period all my direct team members were working well together as a unique team. The customer feedback was positive for some months, although they were unhappy with the advance of their work. They considered their progress should be faster. As a project manager, I talked to the team leader of the consulting company, and we reviewed together the schedule. It was true that they were a little bit delayed, but they argued the customer was not always available to talk to them when they requested to do it. I shared those facts with my customer, and he became confident for some weeks.

However, the deliverables produced by the consulting team were not satisfactory for the customer; they did not see the consultant's value added. They argued that part of my team helped by the customer team could do the consultant's work. Then, the customer asked me to elevate a change request to the project steering committee. The request consisted in firing the team of the consulting company from the project, so my team should absorb the work they were doing in collaboration with customer team members; that news stressed me a lot because the rest of my team members neither me did not have knowledge enough to complete the functional gap analysis they were doing, and neither the new processes implementation. I knew the consultant implementation process methodology, but I only had a period of one month to assign two of my team members, partial time, to learn about the work the consulting team was doing, and asking my customer to collaborate with me to diminish the project impact. The customer accepted and that request for change was approved by the project steering committee, accepting that it would have a schedule impact in the project.

I updated the project plan with my team, and we presented an updated plan to the customer where the project end date would be delayed. For some weekends could not come back to see my family because I was overwhelmed, my two team members assigned to the new tasks needed spend some time with me to learn the process methodology to apply, and it took us some days. We worked over time for some weeks and we progressed, although I felt my team so exhausted. Two weeks later, my family came to visit me, and we could relax a little bit together. My family support, mainly my wife, was key during that period. They understood I had a high responsibility in this project and did not press me.

The positive aspect of that was the project finished, a little bit delayed as expected by all project stakeholders, but successfully, and my customer congratulated us for our team extra effort. When the customer needed us, we were there and did not complain about our extra workload but tried to cooperate with them. It was a satisfaction for me being congratulated by the customer at the end of this project. I learned a lot from this project, during almost two years I needed to face very challenging situations that affected me not only professionally but also personally, but at the end I was well recognized by my customer and my organization.

SUMMARY

In summary, I want to restate the ideas I shared with you in this chapter:

- The critical question is: How often do you complain? If you wonder whether you complain too much, ask your colleagues. They will let you know.
- Over the years, I have noticed that complainers lack perspective; they tend to blow their problems out proportionately.
- Optimistic people tend to understand what is truly important in life. Think about the people you know.
- The people who don't complain very much (and those who speak positively) are a joy to be around. Decide to join that group—so people won't have to cross the street when they see you coming.
- Although your thoughts are not you, if you repeat the same thoughts repeatedly, they will condition your mind to a large extent.
- The mind is like a hyperactive monkey. The more you fight with the monkey, the more hyper it becomes. So, instead, relax and observe the monkey until it wears itself out.

- When you complain, stop and ask yourself if you want to continue denying responsibility for your reality or allow a bit more responsibility back onto your plate.

TOOL—COMPLAINER ASSESSMENT

Please try to answer the following questions and reflect upon my comments for each question.

Constant complaining is so bad because:

1. **It makes things look worse than they are**.

   Are you only focused on what's wrong?

2. **It becomes a habit**.

   Do you feel that everything is terrible, every situation is a problem, every co-worker is a jerk, and nothing is good?

3. **You get what you focus on**.

   Are you focused on negative or positive things?

4. **It leads to one *down* man ship**.

   A complaining session might go something like this:

   The other day, my boss came in five minutes before I left and asked me to finish two massive projects for him. I had to stay two hours and missed my football game.

5. **It makes people despondent**.

   Are you constantly complaining?

   Not only does constant complaining make you see the workplace as worse than it is, but because you're constantly hearing stories of how bad things are and how they're continually getting worse, it destroys all hope that things can get better.

6. **It kills innovation**.

What's the point of coming up with ideas and implementing them—it will never work anyway?

7. **It favors negative people**.

The way to get status among complainers is to be the most negative. To be the one who sees everything in the most negative light.

8. **It promotes bad relationships**.

Are you in a bad relationship with anybody?

People who complain together unite against the world and can create solid internal relationships based on this. However, these relationships are based primarily on negative experiences. That's not healthy.

9. **It creates cliques**.

Are you gathering in cliques with their fellow complainers, where they can be critical and suspicious of everybody else?

Being cheerful, optimistic, and appreciative makes you more open toward others—no matter who they are. Connecting to co-workers in other departments, projects, or divisions becomes easy.

10. **Pessimism is bad for you**.

Are you pessimistic?

Research in positive psychology has shown that people who see the world in a positive light have a long list of advantages, including:
- They live longer.
- They're healthier.
- They have more friends and better social lives.
- They enjoy life more.
- They're more successful at work.

9

Associate with Positive Professionals

You have to perform at a consistently higher level than others. That's the mark of a true professional.

–Joe Paterno

Emile spent much time in high school with many guys in his neighborhood. Emile said these guys just liked to sit on the front porch and watch cars go by. They had no goals and no dreams. They were always negative. Whenever Emile suggested they do something new, the others would discourage him. "It is stupid" or "not cool," they told him. Mike just went along with them to remain part of the group. When Mike went to college, he still encountered some pessimistic people. But he also met positive people who wanted to learn and achieve things. Mike decided to spend his time with positive people. Almost immediately, Mike started to feel much better about himself. He developed a great attitude. He began to set goals. I am happy to tell you Mike now runs his own successful engineering company and has a wonderful family. One by one, he is accomplishing all the goals he has set.

When I asked Mike what happened to his high school friends, he said, "They still live in the same neighborhood. They are still negative. And they are still doing nothing with their lives." Mike added, "I would never be where I am now if I kept hanging out with those guys. I would still be at the corner deli playing pinball." Mike's story is an excellent reminder of the influence others have on our lives. And yet, sometimes, we get in the habit of being with certain people and don't think about the consequences. Have you ever heard the axiom, "Tell me who you hang out with, and I'll tell you who you are?" There is a lot of wisdom in that simple statement.

DOI: 10.1201/9781003485674-9

Have you considered how this principle has been melding and shaping your life?

Think back to when you were growing up. Do you remember your parents' concern about who you hung out with? Your mom or dad wanted to meet your friends and know their details. Why? Your parents knew that you would be significantly influenced by your friends, that you would tend to pick up some of their habits, and that you would probably do what your friends were doing. Your parents were concerned for a good reason. Let me give you an example of how this pattern carries over to professional lives: When I worked for a multinational company as a project manager, I started a PMO (Project or Program Management Office). The professional services organization was not very convinced about the urgency of creating a PMO, so I looked for some allies. Most project management colleagues criticize the lack of support from executives, but they never propose to take any action. One of them thought, like me, he said, "We can do something in terms of our behavior in front of our executives." This colleague became my first ally in the PMO implementation project effort. "Change is possible, and today is a great day" were our preferred sentences.

POSITIVE VERSUS NEGATIVE PEOPLE

Negative people are the ones who always dwell on the negative. Their sentences are contagious; they continually spew their verbal poison (see Figure 9.1). In contrast, positive people promote personal and professional growth; they are very supportive. Positive people lift your spirits and are a gift for all of us. Negative people try to drag you to their level (see Figure 9.1).

May be qualified like energy vampires

Are dream killers

Negative People

Make you feel listless and drained

Will always try to drag you down to their level

Hammer at you with all the things you cannot do and all of the things that are impossible

Barrage you with gloomy statements about the loosy economy

FIGURE 9.1
Negative people.

They hammer away at you with all the things you cannot do and all the impossible things. They barrage you with gloomy statements about the lousy economy, the problems in their life, the issues soon to be in your life, and the terrible prospects for the future. If you are "lucky," they might even say a few words about their aches and pains.

After listening to negative people, you feel listless and drained. I identify some people as "dream killers." I could say that they are "energy vampires" because they suck all the positive energy out of you. Have you ever been with a negative person and felt like that individual was physically taking energy from you? I think we have all had that experience many times. One thing is sure: Spending time with negative people and their messages will wear you down.

On the other hand, how do you feel around positive, enthusiastic, and supportive people? You are energized and inspired (Figure 9.2). There is something genuinely unique about positive people. They seem to have a positive energy that lights up a room. When you are around them, you start to pick up their attitude, and you feel you have the added strength to pursue your goals vigorously.

When I think about positive people, my friend Enrique Capella immediately comes to mind. Whenever I speak with Enrique, I feel I can conquer the world. Enrique is simply the most positive person you could ever meet. I like to think of myself as a very positive person. On a scale of 1–10, with 10 being the most positive, I would probably give myself an 8.5. I would have to provide Enrique with a 14. He is just off the charts. He is always positive, always enthusiastic. And he gives a tremendous lift to everyone who crosses his path. His attitude inspires people to greatness. Can you see how your attitude might improve if you had a friend like Enrique?

Our minds tend to dwell upon whatever we repeat over and over. Unfortunately, the mind does not discriminate between messages that are

Makes you proud to work with him/her		Energyze and inspire you
Makes you happy	**Positive People**	Have a positive energy that lights up a room
Makes you feel you can conquer the world		Makes you feel empowered

FIGURE 9.2
Positive people.

good for us and those that are harmful. If you hear something often enough, you tend to believe it and act upon it. Just as a song repeated many times will get you thinking about that song, so too will repeat thoughts about success get you thinking about success. So, if you fill your mind with positive messages, you will be more positive and move forward boldly to achieve your goals. The more positive reinforcement is, the better. And where can you get this positive reinforcement? Well, one way is to read motivational books. You can listen to motivational tapes and spend lots of time with positive people. I believe that human beings are like sponges: We "soak up" whatever people around us are saying. So, if you spend time with someone negative, you sponge up the negatives, affecting your attitude. Of course, the reverse is also true. When you hang around positive people, you soak up the positive. You feel better and perform better. So, you must join positive people.

ANALYZE YOUR PEOPLE

You must assess your friendships and professional colleagues from time to time, even those you have maintained for many years. Trust me; it is not a minor issue. Those who occupy your time significantly impact your most priceless possession, your mind. Are you surrounding yourself with negative friends and colleagues and spending much time with them in your leisure hours? If so, I will ask you to think about spending much less time with these people, or even no time with them. It sounds harsh. After all, I suggest limiting or eliminating your involvement with some long-standing friends. You can think I am cold or uncaring. You can also feel that we should try to help our negative friends and colleagues instead of dumping them.

Nevertheless, I have found that in most cases, hanging around these negative friends or colleagues does not help them, and it does not help you, either. Everyone gets dragged down because most negative people don't want to change. They want someone to listen to their tales of woe.

If you have a strong urge to spend time with negative people, ask yourself: "Why am I choosing to be with these people?" Consciously or unconsciously, you may be choosing to hold yourself back, to be less than you can become. By the way, trying to help someone overcome their negativity is beautiful. But if you have been pushing for several years and are not getting anywhere, maybe it is time to move on. Let me clarify one important thing. I am not judging that negative people are any less worthy than others. I am saying there are consequences if you spend time with negative people. What are the implications? You will be less happy and less successful than you could be.

One of the things that worked well for me in the projects I managed was when the discussion moved to a negative subject, I resisted the temptation to accuse the other person of being negative. That will usually make things even worse. Instead, gently shift the conversation to a more positive topic. Remember that I am not asking you to disown your relatives or refuse to attend family functions. I am talking about limiting your contacts with hostile relatives, so you do not get dragged down to their level.

DO YOU WORK WITH POSITIVE PEOPLE?

Every organization and every project has some negative people working there. Sometimes, you have to interact and work alongside these people. But do not go out of your way to spend time with these prophets of gloom and doom. For example, if you frequently have lunch with negative people at work, stop having lunch with them. All they are doing is filling your mind with negativity. You cannot perform at your best if you allow these people to dump their harmful garbage into your mind. There is no need to be nasty or to tell them off. You should be able to find a diplomatic way of distancing yourself from this "poisonous" group. See Figure 9.3.

Instead, take charge. Be proactive. Make a point to eat at your desk, take a client out to lunch, or sit at a different table in the cafeteria. Do whatever

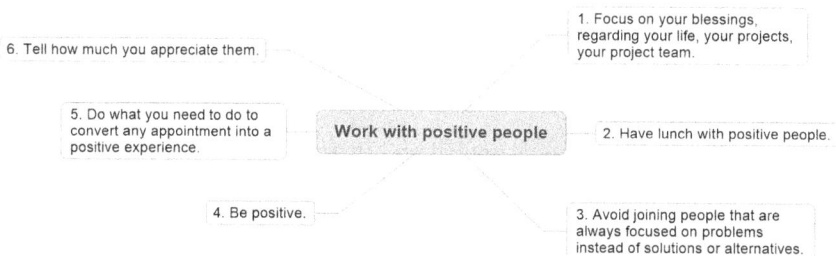

FIGURE 9.3
Work with positive people.

you have to do to make lunch a positive experience. Could you make no mistake about it? Every organization welcomes positive people, and negative people are hurting their chances for advancement. The business community is waking up to the fact that when it comes to productivity in the workplace, attitude is essential for project and business success. Today is a Good Day!

CHOOSE YOUR FRIENDS AND ALLIES

As I said at the beginning of this chapter, "Tell me who you hang out and I'll tell you who you are." Suppose you are serious about getting a raise or a promotion at work, succeeding in your own business, or improving yourself as a human being. In that case, you must start associating with people who can take you to the next level. As you increase your associations with positive people, you will feel better about yourself and have renewed energy to achieve your goals. You will become a more positive, upbeat person, someone others love to be around. I used to think it was necessary to associate with positive people and to limit involvement with negative people. Now, I believe it is essential if you want to be a high achiever and a happy individual. So, surround yourself with positive people. They will lift you the ladder of success.

CASE STUDY 1

After managing essential projects over many years, by 1993, I discovered PMI. I understood the importance of working on my personal development as a project manager. Then, I found different resources, including the PMI Global Congress in the US. After thinking about the benefits for me as a project manager, my peers, my organization, and my customers, I could argue about the need to attend. I hesitated initially, but afterward, I proposed to my manager to hear the Congress. As my performance had been excellent the previous year, he accepted after listening to my arguments. However, he could not understand why a Spanish professional wanted to attend an international congress. He said *Americans are exceptional, and Spanish people are different. I answered him: Thank God we have different cultures, but the point is to be ready to learn from anybody. It will also be an excellent opportunity to meet new people and share similar issues and problems.* It was a fantastic opportunity; I discovered the vast power of networking there. Over the years, I have still met new positive people every year at PM-International Congresses. It encourages me to think about how I manage and improve my projects, generating a lot of positivity and enthusiasm. When I return from Congress, I transmit that enthusiasm to my team in my organization. That's great! I firmly believe there is no border for project management. Positivism is the key to learning from projects and people in organizations.

I continued attending PMI Global Congresses every year and had the opportunity to meet several positive people during my professional life.

I will share with you some examples:

The first is Jim De Piante, PMP consultant and speaker, who is an excellent example of networking. He always has a smile when meeting people and enough charisma to influence people to connect. I was lucky to meet him for the first time in 2006 at the PMI EMEA Global Congress, and since then, we have met each other at many PMI Congresses and events. We delivered some workshops together internationally. He is an example of positivism, looking to connect to new people and share his experience and knowledge.

I also had the opportunity to meet Dr. David Hillson, "the Risk Doctor." I always learn something positive in every conversation. We had colleagues and later good friends. His research and studies about risk management made a difference; he contributed to the project management body of knowledge. He helped me to grow personally and professionally by being constantly positive and giving me direct feedback.

Another example of positivism is Jack Duggal, a project management consultant, speaker, trainer, and good friend. I met Jack for the first time in 1994 in Vienna at the IPMA conference. We met there and had the opportunity to talk because we both delivered a presentation about the subject of PMO. As soon as I spoke to him, I knew I should keep in touch with him. I had the opportunity to work together in several workshops. He is also a smiling guy and is usually very positive.

Dr. Davidson J. Frame, who unfortunately passed away in November 2023, also made a difference in my life with his positive feedback, comments, and help. He was always ready to help me to facilitate my career progress. He was a mentor and a good friend. He supported me when I decided to study for my Ph.D. and encouraged me to learn continuously.

He was another smiling face that I will never forget: a great project manager, academic, author, and speaker who was positively contagious when working with people.

Dr. Michel Thiry also made a difference in my life, demonstrating to me the power of networking in a positive way. He started organizing some dinners in 2003 and maintained them over the years, trying to keep a group of cheerful people collaborating. He created the Valence network, which is still growing. Michel supported me over the years at PMI. I had the opportunity to deliver some workshops with him internationally. We still have a tremendous positive relationship.

My best friend and co-author Randall L. Englund, to whom I need to be grateful for admitting my first HP conference English paper, gave me the benefit of the doubt to deliver my first English presentation. He has made a significant difference in my life, teaching me many things, allowing me to co-write papers, and working together over the years. He introduced me to the PMI Network columns editor in 2006, allowing me to start publishing a column at the PMI Network over the years. Randy always supported me positively in my new initiatives and projects. We became close friends and published some books together. He always maintains his positivism when managing projects and programs and delivering project management consulting and training.

CASE STUDY 2

When working at Hewlett-Packard as a project manager, I found one professional from another department who was very positive (Luis). I had the opportunity to chat with him daily because we had breakfast in the same bar close to the Hewlett-Packard office. He knew of my stress because I always traveled and managed projects outside my city of residence while working at that company. I was married and had three children; my wife also worked but was overloaded with our children, mainly when I was outside the home every week managing a project.

After finishing the most extended project I managed in Spain and managing a PMO in my department, my managers did not recognize me well. I did not get any salary review for two years and decided to leave the organization in one year and create my firm. As I was very confident with him, I shared my plan, and he encouraged me to move forward without talking to anyone else. The first thing I did was start to write my first project management book. Many colleagues considered me arrogant because I wrote a book, but my friend Marcos supported me and was my first reviewer. He gave me some feedback and ideas to improve. At the same time, I started to deliver some training at La Salle University in Madrid, and my efforts were productive. After several months of effort, writing my book, and teaching at the university, when I finished my work at Hewlett-Packard, my deadline for leaving that company was getting closer and closer. Whenever I met my friend, Luis, he encouraged me to move forward and achieve my objective. He introduced me to people who would later help me create the proper computer infrastructure in my new office; he advised me about building relationships with some providers and many more things.

When I announced my resignation one month before my deadline at Hewlett-Packard, most of my colleagues thought I was a fool and doing a hazardous project, but Luis did the opposite. He knew I believed in my dream and wanted to convert it into reality. Some colleagues and managers tried to convince me that my new project would not be successful and that I would fail, but all was negative. On the other hand, Luis was supporting me every day. He was the first person who read my first book. He always reminded me that you must convert your dreams into reality if you believe in them, and I did. I left Hewlett-Packard in April 2002 and got my first big customer in May 2002, managing a significant training program for a savings bank in northern Spain. I continued meeting Luis and celebrating my project's success; he is still a friend, although he retired and moved from Madrid to Malaga. Every time I contact him, he always has a positive thought.

SUMMARY

Remember the following ideas and best experiences about associating with positive professionals:

- Negative people are the ones who always dwell on the negative. Their sentences are contagious; they continually spew their verbal poison. In contrast, positive people promote personal and professional growth; they are very supportive. Positive people lift your spirits and are a gift for all of us.

- You must assess your friendships and professional colleagues from time to time, even those you have maintained for many years.
- Any organization welcomes positive people, and negative people are hurting their chances for advancement. The business community is waking up to the fact that when it comes to productivity in the workplace, attitude is essential for project and business success. Today is a Good Day!

TOOL—ASSOCIATE WITH POSITIVE PROFESSIONALS

Please answer the following questions:

1. Are you joining with positive people?

2. How positive are the people you are in a relationship with?

3. How many positive people?

4. Are around you in your environment?

5. Are you looking for positive people?

6. How much are you smiling every day?

Some suggestions:
If you want to surround yourself with positive professionals, I offer you some tips that might help:

1. **Determine what makes someone happy**: You must determine what makes someone so glad to surround yourself with positive individuals. Optimistic people tend to be upbeat, joyful, and appreciative of their blessings.
2. **Have an upbeat attitude**: Those who are optimistic tend to draw other upbeat people. Therefore, you must be favorable to be among positive individuals. Find something to be grateful for daily, be mindful, and smile even when you're not feeling well.
3. **Join a group or organization**: Making new friends who are passionate about the same things you are is a beautiful way to meet

individuals who share your interests. For instance, you could become a member of Toastmasters International if public speaking interests you.

4. **Volunteer**: Giving back to your community and meeting new people are two excellent reasons to volunteer. People who are passionate about changing the world will be all around you.

5. **Enrolling in a class**: Enrolling in a class is a terrific method to network with others with similar interests. For instance, you could enroll in language, dancing, or cuisine classes.

Remember, surrounding yourself with positive people can positively impact your life. It can help you stay motivated, reduce stress, and improve your well-being.

10

Grow through Your Fears

Even the fear of death is nothing compared to the fear of not having lived authentically and fully.

–Frances Moore

I define myself as a human being always ready to learn something from somebody else; I manage projects, team members, and other project stakeholders. I knew that if you want to be successful, you must be willing to be uncomfortable. To achieve your goals and realize your potential, you must be willing to do things you are afraid to do. That is how you develop your potential. I encourage all project managers I work with to follow this principle. It sounds so simple, but it is not. What most people do when they face a frightening or new situation is to back away from the fear. They do not take action. It is what I did for many years of my life. I firmly believe that is not the correct strategy. Tell me about a successful professional or project manager, and I will show you someone who confronts their fears and takes action; I will tell you about somebody who never loses their courage.

ASSESS YOUR FEARS

Many professionals whom I asked, have you ever been afraid or anxious before trying a new or challenging activity or project? Answered me, "Yes, I was." Most of them told me that sometimes, that fear stopped them from taking action. Sometimes, fear paralyzes people. I remember the first project I managed: *I was twenty-seven years old, without experience as a project*

DOI: 10.1201/9781003485674-10

leader, and I had to share the project mission, objectives, scope, initial risks, and so on with a large group of executives. I was very nervous; it was the first time I was in front of a big group of people who were my project stake-holders. When I started to present, my mind stopped and paralyzed for a few seconds. However, as soon as I started talking, I felt more comfortable. I learned those reactions are part of us as human beings. Of course, every person has a different fear threshold. What frightens one person to death might have little impact on someone else. For example, speaking in public or starting a new business seemed scary some years ago. Others might be fearful about asking someone for directions or deadlines. This lesson applies to you regardless of how trivial or silly you believe your fears may be. When I talk about fear, I am speaking about those challenges that stand in the way of your personal and professional growth. These are the things that scare you but which you know are necessary if you are going to get what you want in life.

THE COMFORTABLE ZONE

Every time you step out of your comfort zone, you will increase your anxiety because of fear (see Figure 10.1). Each of us has a comfort zone: a zone of behavior that is natural and familiar to us and where we feel comfortable and safe.

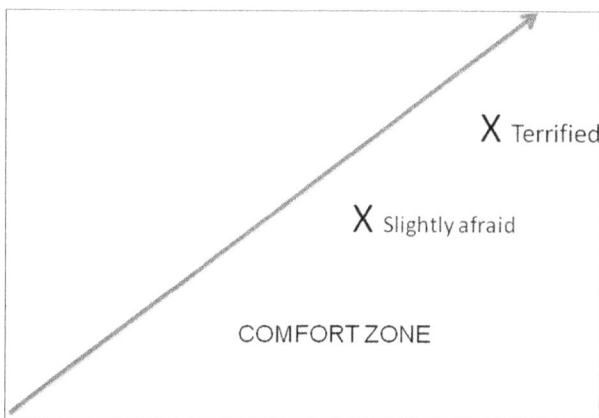

FIGURE 10.1
Comfortable zone.

FIGURE 10.2
Your fears as a professional.

The activities and situations inside the comfort area are non-threatening and familiar. They are routine, part of your everyday life, the things you can do with no sweat. In this category are tasks such as speaking to your friends, colleagues, or customers or filling out the daily paperwork at your job. However, you face experiences or challenges outside your comfort zone as a project manager. These are represented by the "Xs" in the diagram above. The farther the "X" is from the comfort zone, the more afraid you are to participate in that activity. When faced with something outside your comfort zone, you suddenly feel nervous. Your palms become sweaty, and your heart pounds. You begin to wonder, "Will I be able to handle it? Will others laugh at me? "What will my colleagues and customers say?"

Looking at the diagram above, what does the "X" represent for you? In other words, what fear is holding you back from reaching the next level of success or fulfillment in your life?

Whatever that "X" represents for you, admit it honestly. I guess thousands, if not millions, of people have the same fears as you. Let's take a closer look at what most project managers fear. See your fears as a professional in Figure 10.2.

THE MOST COMMON FEARS

I have asked many project manager professionals about their common fears as project managers manage organizational projects (see Figure 10.3). From professionals across different countries and cultures, the same answers come up again and again. Here are some of the most common fears they identify:

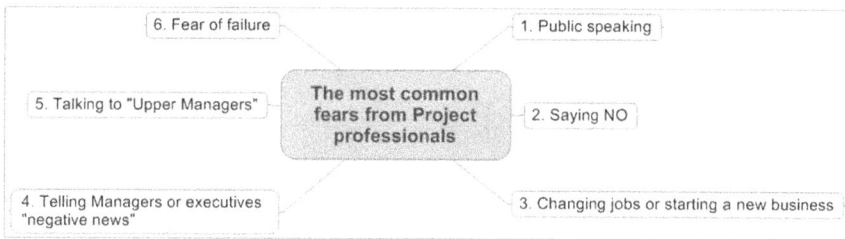

FIGURE 10.3
The most common fears of project professionals.

Are you surprised by any of the fears on this list? I believe most project professionals experience these fears at some point in their projects. And if you have some fears that were not on the list, do not worry about them; TODAY IS A GOOD DAY! You are stronger than any of your worries and can overcome them.

BACKING AWAY FROM YOUR FEARS

When confronted with an anxiety-producing event, most people will retreat to avoid the fear and anxiety. That's what I used to do. You see, backing away does relieve the fear and anxiety that would have resulted if you followed through with the activity. For example, suppose someone asks you to make a presentation within your organization, and you decline. In that case, you save yourself the sleepless nights you would have worried about it and the nervousness you would have experienced in the days leading up to the presentation. I have found that is the only benefit you get by retreating—a momentary avoidance of anxiety. Think about it for a moment. Do you know any other benefits people get when they refuse to confront their fears? Nobody has been able to tell me any additional benefits. I believe there are none.

THE PRICE YOU PAY

Now, I want you to seriously consider the price you pay when you back away from fears standing in the way of your growth. Here is what happens; see Figure 10.4.

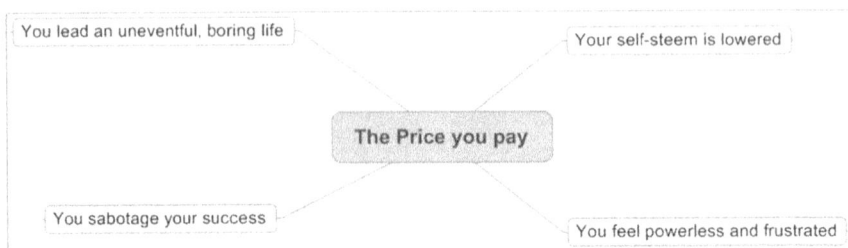

FIGURE 10.4
The price you pay.

Many of us are willing to pay this dear price to avoid temporary discomfort and possible ridicule from others. I think it is not good. Returning is not the best way to handle your problem in the long run. You will never be highly successful or fully develop your talents unless you are willing to confront your fears.

STRATEGY

I was pretty shy in high school and didn't feel good about myself. But I was never rejected when asking someone for a date. If you were looking at me now, you would probably think, "Alfonso is not bad-looking, but he is not Robert Redford." My strategy was quite simple. I never asked anyone out on a date. You see, I was not going to let anyone reject me. And what did I accomplish? I felt horrible about myself. I knew that I had "wimped out." I felt powerless; as you can imagine, I didn't have a full social calendar. I was sabotaging my success. Because I refused to face my fear, I remained in the background while most of my friends and classmates went out on dates. How do you think that made me feel? It's pretty lousy, just as you would expect. In case you are wondering, I had a few dates during that period, but only when other people arranged them. I would not allow anyone to say "NO" to me. In reality, I was saying "NO" to myself. Can you see how my strategy of backing away from my fears worked against me? Now, it is true that if I had asked some people for a date in high school, a few of them might have said "NO." But you know what? I would not have died. I could have asked another person and another, and eventually, I would have gotten a "YES." It was not until college that I began to take some "baby steps" to confront this fear of rejection. Little by little, I gained more confidence. While studying Computer Science, I had the good fortune to meet Rose, and we have been married for thirty nine years. That situation happened to me during my career as a project manager. I had some opposing managers and sponsors while managing projects in organizations. However, I gained more and more confidence as I became engaged in multinational projects and activities. And people involved in those projects welcomed my expressions of leadership, not rejection. I felt much more recognized and satisfied. It was when I discovered the incredible power of my positive behavior in dealing with people. That insight changed my life completely. I started talking to my executives and project sponsors as human beings first, then as professionals. I lost some of my fears in front of them. Little by little, I started to feel more comfortable. My worries were there, but I felt more and more comfortable dealing with them. It was great.

A NEW LIFE

I am no different from you. I have my fears, just as you do. And when I look back at the first thirty years of my life, you know what I see? I see someone who achieved some degree of success as an engineer. But I also see someone shy, insecure, scared, and self-conscious. Does that sound to you like someone who is a motivational professional? What turned my life around and improved it a million-fold is that I learned to confront my fears and take action. After years of frustration and disappointment, I realized that hiding from my worries was not getting me anywhere and would never get me anywhere. Of course, I would not have confronted my fears if I had not first developed a positive attitude. A "can-do" attitude gave me the extra push to act. You dare to move forward despite fear when you believe you can do something. Armed with a great attitude, I decided to participate in life and explore my potential despite fear. From the very beginning, I felt so much better about myself. I had taken control of my life, and many possibilities opened up. Are you beginning to see the incredible rewards you can receive when you are willing to develop a positive attitude and confront your fears?

REFRAME THE SITUATION

If I could give you a way to confront uncomfortable situations without fear or anxiety, you would be ecstatic and eternally grateful. Sorry, but there is no such magical solution. I cannot wave a magic wand and take away your fears. How can you muster the courage to do things you fear but need for your success and growth? The next time you face a scary situation, I suggest you take a different outlook. Most people start thinking, "I will not be able to do this well, and others may laugh at me or reject me." They get hung up about how well they are going to perform. Because of these worries, they decide to retreat. Choose to think differently. While you should always go in with a positive attitude and prepare beforehand to the extent possible, don't be overly concerned with the result. Consider yourself an immediate winner when you take the step and do the thing you fear. That is right. You are a winner by participating in the arena, regardless of the result.

MOVING FORWARD

For example, let's assume you are afraid to speak in public, but you confront your fear and do it anyway. The moment you get up and speak before the audience, you are a winner. Your knees may be shaking, and your voice may be quivering. That does not matter. You faced your fear and accepted the challenge. Congratulations are in order. The likely result is that your self-esteem will enhance, and you will feel exhilarated. Use the fact that most audiences want the speaker to succeed to your advantage.

CASE STUDY 1

If we transported ourselves back in time to, say, 1700 A.D., we would likely find that the medicine men and women in traditional cultures were keen observers of nature, able to interpret weather patterns, the behavior of birds, animals, and other things necessary for survival. After all, they spent their entire lives in close interaction with the natural landscape and received the gift of wisdom and a deep understanding of the cycles of nature. Those men and women were tribal psychiatrists and caretakers of ancient medical knowledge. As such, they asked them for advice, and those men and women listened to their words carefully. In some instances, the shaman would supervise the vision quests of young people on the cusp of adulthood and about to embark upon adult lives as valuable members of society. While on the vision quest, these young people sought the guidance and aid of spirit allies. The shaman would interpret the results of the quest and give meaning to things that were otherwise difficult to understand. The vision quest was also a means of facing real or imagined fear because it took great courage to go off alone, fast, and endure various perils. In effect, the quest was a means of confronting fear, while the shaman was there to advise and direct the participant and help the person overcome the fear of the unknown and the spirit world.

To face and confront one's fears or other obstacles in life is to gain an increasing measure of freedom. The result is that external influences or internal habits can shape your life less. You become increasingly open to change, to a willingness to move in directions that change your circumstances for the better. In other words, fear loses its grip on your behavior patterns and how you live your life.

CASE STUDY 2

I always perceived my boss as my "father," so I had an obligation to obey them in any circumstance of my job. The justification for my behavior was my autocratic childhood education, where the figure of the father meant total respect and absolute obedience. During my professional life, I dealt with my bosses in the same way; I thought my manager was like my father but in a very autocratic way. I could not have a different opinion about any subject; my manager was suitable for everything, and I never thought I needed to have my own opinion and criteria. My behavior generated a lot of stress when I started working as a project manager, and my fear of my manager had to disappear over time. I understood that my manager was also a human being like me, with different responsibilities and metrics. Still, I understood I needed to work with my manager, listen to them, and give my opinion based on my experience and results. I need to say that I grew from my fears, and several years later, I was in charge of executives training on project sponsorship. So, from then on, I have been dealing with executives and collaborating with them to achieve more successful projects. Working together as a binomial, I had the opportunity to learn the language executives understand and educate my peers and colleagues about it. Learning from executives and linking projects to the organizational strategy have been crucial for my professional career. I firmly believe every project manager can grow from their fears if they move forward and try to obtain some positive results, but you also need to apply persistence to be successful. Manage your fears and grow!

SUMMARY

- When facing a frightening or new situation, most people do not return from the fear. They do not take action. It is what I did for many years of my life. I firmly believe that is not the correct strategy.
- When I talk about fear, I am speaking about those challenges that stand in the way of your personal and professional growth. These are the things that scare you but which you know are necessary if you are going to get what you want in life.
- When confronted with an anxiety-producing event, most people will retreat to avoid the fear and anxiety. That's what I used to do. You see, backing away does relieve the fear and anxiety that would have resulted if you followed through with the activity.
- I am no different from you. I have my fears, just as you do. And when I look back at the first thirty years of my life, you know what I see? I see someone who achieved some degree of success as an engineer. But I also see someone shy, insecure, scared, and self-conscious.
- Learn to think differently, take action, and grow through your fears.

TOOL—ASSESS YOUR FEARS

Please answer the questions as follows and reflect on them.

1. **Do you have any fear?**
   Understand what is real versus your perception of the feeling. (Our perception creates the world as we see it, and everything we perceive

is our reality). However, we need to be aware of our biases, judgments, and expectations, and how we see situations is not always based on evidence but on our fears or what could happen.

2. **Take a moment to understand the trigger**.
   "An emotional trigger is anything—including memories, experiences, or events—that sparks an intense emotional reaction, regardless of your current mood."

3. **Ask yourself where the fear is in your body**.
   A lot of times, fear takes over physically. It affects different people in different ways. Some people start to fear in their stomachs, or maybe they get a headache. Remember that your emotions always come out in your body when you're out of balance.

4. **Practice gratitude**.
   Every day, list out one to three things you are thankful for. It doesn't matter how big or small it is; gratitude helps shift the mind into a positive light, which, over time, overcomes fear. Are you grateful for the people in your life? Your health? Your job? Your home? Safety? Sometimes, it can be hard to find things to be grateful for if you're in a difficult spot, but you always have things to be thankful for.

5. **Listen to your inner voice**.
   Monitor your inner conversations. We often say mean things to ourselves, tear ourselves down, and beat ourselves up without realizing it. Would you criticize a friend for messing up or tell them they're going to eff it all up before an essential challenge in their life? Or would you tell them: "You've got this!"? When you find yourself doing this and going through all the WHAT IF'S that could happen and creating a scary scenario, go through those thoughts and answer yourself with:
   - What if I do get what I want?
   - What if I succeed?
   - What if I get the promotion?
   - What if they talk to me?

6. **Replace with a new association**.
   When you start to feel fear in your body that is creating discomfort and anxiety, remind yourself that the feeling and the moment will pass. Instead of walking unthinkingly into a stressful situation or one that could cause you to freeze up, try preparing in advance for problems you know might cause you fears or self-doubt. If it's a specific situation you know causes fear, or something where you

anticipate the anxiety ahead of time in the examples of an important presentation, an interview, a difficult conversation, putting yourself out there and going on a date, asking for a promotion, etc., please think that situation will be temporary and it will end.

7. **Shift to a positive mindset instead of dwelling**.
   Perception is compelling, and how you feel about your situation dictates your response. It's normal when we think we mess up for us to go over it over and over in our heads. Maybe we beat ourselves up in our heads and scold ourselves for past mistakes and how we messed up. We've all probably done it—replaying something you feel like you effed up over and over.

8. **Practice breathing exercises, tapping to rebalance**.
   Breathing helps center your body; when you stop breathing, your heart stops beating. You can do a grounding exercise or even take five deep, long breaths at any point to calm and center yourself. It is best to start your day with this but feel free to practice all day. Many people find box breathing helpful.

Here are the simple instructions for this potent exercise:

Breathing describes the pattern of inhaling slowly and deeply through your nose to the count of four. You inhale for four seconds, hold for four seconds, exhale for four seconds, and then repeat in four seconds—making a square pattern.

11

Get Out There and Fail

You always pass failure on the way to success.

–Mickey Rooney (1920)

I remember my father's words long ago: *If you want to learn to do something well, you must fail first.* I believe failing is an excellent way to remember to do things better. Henry Ford said: *Failure is only the opportunity to begin again more intelligently.* One of the lessons I learned as a project manager was that persistence is the key to project success. I failed many times in my professional life but was willing to keep failing until I succeeded. Alexander Graham Bell said:

> What is power is I cannot say; all I know is that it exists and it becomes available only when a man is in that state of mind in which he knows what he wants and is fully determined not to quit until he finds it.

Human beings do projects and make right and wrong decisions during the project life cycle. This fact is part of human behavior that many executives forget exists during project execution in organizations. How to learn from successes and failures characterizes a project learning organization. One of the obligations of executives is to plan with their project managers to do retrospective analysis during each project's life cycle. If lessons-learned sessions are not part of the project plan, they never will happen.

Trying hard and making many attempts is known as commitment and persistence in a general sense. Project teams and managers achieve right or wrong results, considering that projects are uncertain. Depending on

DOI: 10.1201/9781003485674-11

the point of view of those teams and leaders, those results will appear as failures or opportunities to learn. For instance, Thomas A. Edison "failed" more than five thousand times before inventing the incandescent light bulb. However, when some people talked to him about these "failures," he said: "I did not fail; I discovered five thousand ways of how not to build an incandescent light bulb." The learning attitude is a good characteristic of the right project leader. Every day, I can learn something in my project; it does not matter if I know from my people, customers, or other project stakeholders. Good ideas or feedback can come from anywhere. The most important thing is how to move from failures to project success.

SOME YEARS AGO

I remember when I was a child and learned how to ride a bicycle. Perhaps you had a similar experience that began with training wheels. Keeping your balance becomes more complicated when you remove these crutches. You struggled to stay upright, maybe even falling a few times and scraping yourself. You were learning an important early lesson about failure. As you practiced, one of your parents likely walked beside you, shouting instructions, encouraging you, and catching you as you lost balance. You were scared but excited. You looked forward to when you would succeed and, finally, ride free. Or maybe you didn't think but were wrapped up in the experience and how to accomplish the activity. Nobody called you a failure, nor were you worried about failing. So, you kept at it every day and eventually mastered riding a bike.

What contributed to your ultimate success in learning how to ride your bike?

See Figure 11.1.

Well, persistence and sheer repetition, indeed. You were going to stick with it no matter how long it took. It also helped that you were enthusiastic about what you set out to achieve and could hardly wait to reach your goal. And finally, let's not underestimate the impact of positive encouragement. You always knew your parents were in your corner, supporting and rooting for your success. As a youngster learning to ride your bike, you were

FIGURE 11.1
Contributors to your ultimate success.

optimistic, thrilled, and eager to meet the challenge. You could not wait to try again. You knew you would master it eventually. But that was a long time ago.

YESTERDAY AND TODAY

Now, let's examine how most adults approach the development of new skills. Let's assume we asked a group of adults to learn a new software program or switch to another company position. How would most respond? They would try to avoid it, complain, make excuses because they should not have to do it, doubt their abilities, and be afraid. As adults, most of us become a lot more concerned about the opinions of others, often hesitating because people may laugh at us or criticize us. As a youngster, we knew we had to fall off the bike and get back on to learn a new skill. Dropping off the bike was not a "bad" thing. But as we got older, we started to perceive falling off as a bad thing rather than an essential part of achieving our goal. It can be uncomfortable to try something new, perhaps even scary. But if you take your eyes off the goal and instead focus on how others may view you, you are doing yourself a grave disservice. To develop a new skill or reach a meaningful target, you must be committed to getting there, even if it means putting up with negative feedback or falling on your face now and then. Successful people have learned to "fail" their way to success. While they may not enjoy their "failures," they recognize them as a necessary part of victory. After all, becoming proficient at any skill requires time, effort, discipline, and the willingness to persevere through difficulties. Persistence is the key.

THE GREATEST MISTAKE

When I ask you to name the best basketball player of all time, who comes to mind? I am guessing that many of you immediately thought of Michael Jordan. He gets my vote. Let me share this statistic with you: Michael Jordan has a career shooting percentage of fifty percent. In other words, half of the slots he took in his career were "failures." Of course, this principle is not limited to sports. We also know that show business stars and media personalities are no strangers to failure. I spent many years learning the basics of project management, and I still make some mistakes in the projects I manage.

However, I try to learn from them. I recognize my failures in front of my people. I am not Superman; I make mistakes every day, but at the end of the day, I try to summarize my mistakes, promise to make adjustments, and plan that the next day will be great. You will succeed if you keep trying, developing yourself, and making adjustments. You must get enough at-bats, attend enough auditions, and visit enough potential clients. Whenever I make a mistake managing a project, I recognize it and say, "I made a mistake, and I will do my best to correct my mistake; my apologies." It is the type of behavior I get across to my people. This attitude is constructive in my life as a project manager. So, the greatest mistake of a project manager is not recognizing that something went wrong and not saying, "I made a mistake."

THE POWER OF PERSISTENCE

I believe commitment is the essence of a learning attitude. The key to getting what you want is the willingness to do "whatever it takes" to accomplish your objective. What do I mean by this willingness? It is a mental attitude that says: If it takes five steps to reach my goal, I'll take those five steps, but if it takes thirty steps to achieve my goal, I will be persistent and take those thirty steps. On most occasions in the project field, you don't know how many steps you must take to reach your goal or to accomplish your deliverables. The number of steps to carry does not matter. To succeed, all that's necessary is to commit to whatever it takes, regardless of the number of steps or activities involved. Persistent action follows commitment. It would help if you were committed to something before you persist in achieving it. Once you save to reach your goal, you will follow through with relentless determination and action until you attain the desired result. The most challenging thing I found was convincing the team about the significant impact on business that commitment within projects has on organizations. When you commit and are willing to do whatever it takes, including the effort to communicate a clear, convincing, and compelling message, you begin to attract the people and circumstances necessary to accomplish your goal.

For instance, some years ago, I was responsible for implementing a Project Management Office at an IT multinational company. The manager of that organization did not believe in project management. My team consisted of junior people without experience in project management. However, I achieved tangible results in a few months. The key to project success was that my vision "turned from potential failure to possible success" in my mind, and it was an incredible engine. I spent a lot of time meeting project stakeholders and searching for project allies in the organization, meeting team members, explaining the purpose and objectives of the Project Office, and training them on project management basics. In this story, the key to project success was explaining that everyone must be accountable and responsible for their tasks and activities; that means getting team members to commit. Speaking the truth about the project to all project stakeholders was also crucial. I created a need to learn the project life cycle, identifying right and wrong insights. I believe that once you commit yourself to something, you make a mental picture of what it would be like to achieve it. Then, your mind immediately goes to work, attracting events, circumstances, and people that help bring your picture into reality.

It is important to realize, however, that this is not a quick process, and you need to be persistent and, most importantly, educate and support your people about persistence. I believe that is a must for project success. Usually, project failures precede project success, but success will come, especially if all project stakeholders are ready to learn, reflect, and take action toward achieving desired results.

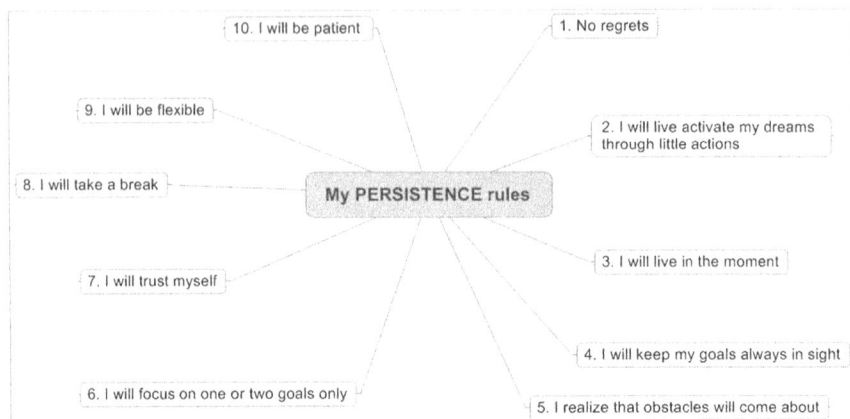

FIGURE 11.2
My persistence rules.

MY RULES OF PERSISTENCE

Therefore, today, I have decided on some "rules of persistence for my projects and myself." See Figure 11.2.

1. No regrets. I will follow my dreams to the fullest. With all my energy, I will give it my complete will and effort. So that even if the desired result does not come about, I will have no regrets. I know I tried. I honor all decisions from my freedom to choose and accept the consequences.
2. I will live and activate my dreams through little actions. Yard by yard, push by little push. I need not take massive action each day. But a little measurable step forward will bring me that much nearer to my goal.
3. I will live in the moment, not in the past, and not too much into the future. The full realization of the present will make the future come about independently. Let me focus all my energies on the present, so I don't rue the time lost when the clock ticks over.
4. I will always keep my goals in sight. I have written down goals. I carry a copy with me in my wallet. I am making a daily habit of going through it at least once. The other place I have kept it is as wallpaper on my monitor. It's always in my face and in my subconscious, too.

5. I realize that obstacles will come about. I need to work around them. Goals are what lies behind all the stumbling blocks. If I cannot vault over them, I will walk around them. It might take longer, but I will get around the block.

6. I will focus on one or two goals only. Focus is concentration on one single point. It's much easier to be persistent when we have clarity of a single goal. Too many goals dissipate our energies; when you lose power, you forfeit your persistence.

7. I will trust myself. When others can do it, so can I. I try out this mantra every day. I know all the power to achieve my goals lies within me. I only have to harness it.

8. I will take a break. I have to fill myself up with energy. After every slight success, it's important to taste a reward. To chill out for a while and then get back on the job rejuvenated.

9. I will be flexible. Constant action sometimes demands inconsistent methods. If a way is not working well, I will try to find another way. There is always more than one way to skin the cat.

10. I will be patient. What defeats persistence is time. Time is our greatest friend as well as our greatest enemy. Persistent action, by its very extension, means overcoming an obstacle over time. So, I must make time, my ally, and trust that I will complete my goal with patience. I will have utilized time if I can progress a little each day.

NEVER GIVE UP

In 2002, the IT manager of a savings bank called our office to inquire about our project management services, as well as our products and publications. When we called to follow up, he said he was "thinking about it and had not made a decision yet." In the beginning, we called every week. No sale. Then, we called once a month. No sale. For a few years, we kept calling this gentleman. We kept sending him quarterly newsletters and flyers. And all we had to show for it was one failure after another. But in the autumn of 2004, a representative of his company called our office, and they hired us to deliver a consulting and training program for more than 150 employees. When I met the IT manager, he said, "I was impressed with your persistence. Someone from your office kept calling me for years and did not give up." Sure, we put up with years of failure. But it was all worth it when we made that sale.

Let me share with you another example. In 2007, PMI Madrid members elected me as PMI Madrid Chapter President. PMI Madrid had about 250 members and lacked activities and few success stories. I am still in a challenging situation, and some board members resigned. However, we were financially healthy and grew to 439 members that year. While voluntarism is essential for this association, it isn't easy to come by in Spain. Moving forward is difficult but not impossible; TODAY IS A GOOD DAY. I will continue working hard and trying to motivate and encourage my team and our members.

IMPORTANT QUESTIONS

If you are not getting the results you want or have been discouraged by failures, ask yourself these questions:

1. Do I have an unrealistic timetable? Maybe you expect to "skip steps" and succeed on a grand scale immediately. Success accomplishment happens by climbing one step at a time. And you do not always know how long it will take to advance to the next level. So, be patient with yourself—and resist the temptation to compare your progress with anyone else's. You will grow faster than some and slower than others. Maintain a great attitude, take action, make adjustments, and the results will come.

2. Am I truly committed? It is essential for me that you be willing to do whatever it takes and banish any thought of giving up before you accomplish your objective. Of course, your commitment is much easier when you love what you are doing. Therefore, go after those goals you are passionate about and harbor no thought of quitting.

3. Do I have too many discouraging influences? Unsuccessful results can be frustrating. That is why we must surround ourselves with people who support and believe in us. Suppose you hang around with negative people who are highly critical or doing very little in their lives; your energy and enthusiasm drain. Therefore, develop a network of individuals to encourage and coach you toward success.

4. Am I preparing to succeed? Success in any endeavor requires thorough preparation. Are you taking steps to learn everything you can about accomplishing your goal? Preparation means reading books, listening to tapes, taking courses, and networking with highly

successful people in your field. It might mean finding a mentor or getting a coach to work with you. Successful individuals are always sharpening their skills. Those getting unsuccessful outcomes do the same things repeatedly without making necessary adjustments. So, be "coachable." Accept that you do not know it all and find resources to keep you on track and moving forward.

5. Am I truly willing to fail? Face it, failure is inevitable. You will encounter defeat before succeeding. In our hearts, we know our most valuable lessons come from our failures. Failure is essential for growth. Look failure squarely in the face and see it as a natural part of the success process. Then, failure will lose its power over you. When you are not afraid to fail, you are well on the way to success— welcome failure as an unavoidable yet vital component in achieving your goals. Be tolerant of failure in others and the organization, and encourage learning from those failures.

TURNING FAILURE INTO SUCCESS

Your failures are learning experiences that point out the adjustments you must make. Never try to hide from failure, for that approach guarantees that you will take virtually no risks and achieve very little. Singer Beverly Sills once remarked, "You may be disappointed if you fail, but you are doomed if you do not try." No, you will not close every sale. And you will not make money on every investment. Life is a series of wins and losses, even for the most successful. The winners in life know that you crawl before you walk and walk before you run. And with each new goal comes a new set of failures. Whether you treat these disappointments as temporary setbacks and challenges to overcome or as insurmountable obstacles is up to you; if you make it your business to learn from every defeat and stay focused on the result you wish to attain, failure will eventually lead you to success.

CASE STUDY 1

The Maine Medicaid Claims System project is a case study of an awry project. The project aimed to switch from their legacy systems to a new web-based system to process Medicaid claims and facilitate HIPAA compliance

(Health Insurance Portability and Accountability Act of 1996). As a result of the failed project, Maine is now the only state in the union not in compliance with HIPAA. System problems led to many claims ending up in limbo, leading to hundreds of calls from health care practitioners, turning away nearly 300,000 patients, several dentists and therapists going out of business, and destroying Maine's finances and credit rating. So what went wrong?

Mistakes included the following:

- Deciding to develop an entire system from scratch using unproven technology while other states built a front end onto their legacy systems
- Caving to pressure from management to meet tight deadlines with inadequate resources instead of pushing for a realistic plan to begin with
- Failing to notice why other bidders either didn't bid or came in way higher (a sign that the schedule was unrealistic)
- Hiring a vendor with no experience developing Medicaid claims systems because they were the lowest bidder
- Not having a Medicaid expert on the team leads to errors in judgment
- Underestimating the time needed to meet with subject matter experts
- Competing with another significant initiative (a department merger) for executives' attention and resources
- Skipping project management basics (including piloting, adequate end-to-end testing, staff, and user training) due to looming deadline pressures
- Failing to stop, regroup, and analyze the risks
- Taking a "big bang" approach to cutover with no contingency or backup should something go wrong

CASE STUDY 2

At the beginning of my project management career, my organization assigned me the role of project manager for a software development project. It was an IT project to implement one software application in a Spanish bank. I had a team of six people (software developers), and my customer expected to have the system up and running in ten months. I was a junior project manager with only a few years of experience managing projects. On average, people in my team were ten years older than me. I was an accidental project manager. I remember that I could not sleep well by night because I was apprehensive about managing my team; my previous experience was working with another colleague on a simple project. Our project was more complicated and required an excellent leader to move forward and motivate the project team. As I was inexperienced, I first talked to everyone on my team and asked for help. I used the sentence "I want you, and I need you," showing every team member that I needed their cooperation and collaboration to succeed. Initially, it was a challenge, but after a couple of months, my team members appreciated how I dealt with them, the respect I used with them in my interactions, and how I defended their work in front of the customer. I knew it was my first significant project but I failed some tasks. But I learned, and my team members supported me. I had a lot of fear at the beginning; I was alone in my room many evenings, depressed because I was outside of the home and far away from my family, but step by step, my relationship with my team members improved dramatically. We were in a similar situation and wanted to finish our project successfully. My team members knew their jobs very well; in my case, I learned how to prepare a plan and coordinate the teamwork, but I failed in my interaction with the team. My team members taught me that together, we could be successful. I needed to fail initially to be aware of my mistakes and learn from each other. I understood that everyone in my team could be my teacher in some aspects. I knew how complementary team members must be in the way of project success.

SUMMARY

- Human beings do projects and make right and wrong decisions during the project life cycle. This fact is part of human behavior that many executives forget exists during project execution in organizations.
- Good ideas or feedback can come from anywhere. The most important thing is how to move from failures to project success.

- It can be uncomfortable to try something new, perhaps even scary. But if you take your eyes off the goal and instead you need to focus on how others may view you, you are doing yourself a grave disservice.
- Whenever I make a mistake managing a project, I recognize it and say, "I made a mistake, and I will do my best to correct my mistake; my apologies." It is the type of behavior I demonstrate to people.
- I believe commitment is the essence of a learning attitude. The key to getting what you want is the willingness to do "whatever it takes" to accomplish your objective.
- It is important to realize, however, that this is not a quick process, and you need to be persistent and, most importantly, educate and support your people about persistence. I believe that is a must for project success.

TOOL—PERSISTENCE ASSESSMENT

Please select your option and reflect upon it to prepare your improvement plan.

Persistence despite difficulty:

1. I keep on going when the going gets tough.
2. People describe me as someone who can stick to a task even when it gets difficult.
3. Even if it is difficult to understand, I will read an entire book until I "get" it.
4. Setbacks do not discourage me.
5. Even if something is hard, I will keep trying at it.

Persistence despite fear:

1. I tend to face my fears.
2. Even if I feel terrified, I will stay in that situation until I do what I must.
3. I stay persistent even when I am scared of things.
4. If I am worried or anxious about something, I will do or face it anyway.
5. If something is scary, I will do it anyway.

Inappropriate persistence:

1. I prefer to work on long-term goals.
2. Most of the goals I work on take years to finish.
3. I usually work toward small goals.
4. I like to work on short-term goals.
5. Most goals I accomplish only take a few days to complete.

12

Networking

Everyone is born with genius, but most people only keep it for a few minutes.

–Edgard Varese

The sooner you start creating a network, the faster you will progress in your career. An incredible chain reaction began when I joined Project Management Institute (PMI) in March 1993. That would forever impact my professional life and then my current business. Let me share with you what happened. At the end of 1992, I attended project management training in France, organized by HP, where I worked for almost fourteen years. In that training, the teacher distributed some project management articles to the attendees, and I learned about the existence of PMI as a professional association. I asked to attend the PMI Global Congress in 1993 to my manager at HP, and after some discussions, he accepted my request. A vast window opened for me when I went there.

On the first day of the Congress, I was slightly frustrated because I was the only Spanish professional attending that Congress. I was conscious that many project management practitioners in Spain were not there. I participated in a session called Global Forum, organized by Mr. David Pells, and there, I met most professionals I would be in relationships with some years later. I had the opportunity to distribute many business cards and collected many from colleagues from different countries and areas of expertise. I had a good time talking to and connecting with people. That compelling first event motivated me, and I understood the vast

DOI: 10.1201/9781003485674-12

power of networking with people. Over the years, I continued attending those annual Project Management Congresses, and I had an extensive network that was increasing yearly. The power of networking is nothing short of incredible. If given the choice, wouldn't you like to succeed sooner rather than later? Well, networking is a way to leverage your efforts and accelerate the pace at which you get results. I firmly believe that the more solid relationships you build, the greater your opportunities for success.

THE GREAT BENEFITS OF NETWORKING

I believe your success starts with you; however, it grows to higher levels due to your associations and relationships with people. Simply put, you cannot succeed on a grand scale alone. That is why networking is so important. I can define networking as developing relationships with people for mutual benefit. I always took care of keeping my network alive. You can see some of the business benefits you can get as a project professional from networking activities in Figure 12.1.

But what can we do to enhance the effectiveness of our network? I found some productive techniques that have been very helpful for me. I have classified it into different categories: taking action, references, communication, and follow-up.

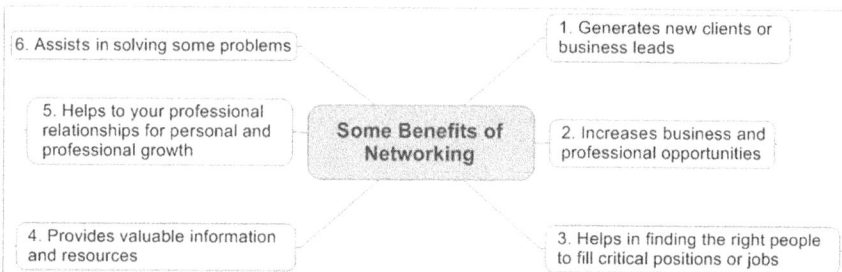

6. Assists in solving some problems

5. Helps to your professional relationships for personal and professional growth

4. Provides valuable information and resources

Some Benefits of Networking

1. Generates new clients or business leads

2. Increases business and professional opportunities

3. Helps in finding the right people to fill critical positions or jobs

FIGURE 12.1
Some benefits of networking.

TAKING ACTION

1. *It would help to project a winning attitude*: When discussing net-working, attitude is a key to success. People will want to spend time with you if you are positive and enthusiastic. They will want to help you. If you are gloomy and pessimistic, people will avoid you and hesitate to refer you to their colleagues.

2. *Be active in organizations and associations*: Effective networking and relationship building takes more than paying dues, putting your name in a directory, and showing up for meetings. You must demonstrate that you will accept the time and make an effort to contribute to the group. What kinds of things can you do? For starters, you can volunteer for committees or serve as an officer or member of the board of directors. The other members will respect you when they see you roll up your sleeves and do some work. They will also learn about your people skills, your character, your values, and last but not least, your attitude to accept gladly.

    Let me give you the example of Francisco. He wanted to build contacts within the electronics market, so he joined the Electronics Firms Association in Madrid in 1998. Francisco immediately began to attend the group's meetings. When they asked for volunteers for various projects, Francisco raised his hand. He got actively involved. Within six months after joining, somebody approached him and said. "We hear good things about you. You are a hard worker and very energetic. Would you like to join our board of directors?" As you might guess, Francisco gladly accepted. Within a few months, he saw a significant increase in his business. In early 1999, Francisco told me that well over fifty percent of his current business traces to people he met through the Electronics Association, proving that people can quickly get significant results by networking.

3. *Serve others in your network*: Serving others is crucial to building and benefiting from your network. You should always be thinking, "How can I serve others?" instead of "What's in it for me?" if you come across as desperate or as a "taker" rather than a "giver," you will not find people willing to help you. Going the extra mile for others is the best way to get the flow of good things coming back to you. How can you serve others in your network? Start by referring business leads or potential customers. In addition, whenever you see

an article or other information that might interest someone in your network, forward the material to that person.

When I think of effective networkers, the first name that comes to mind is Jim De Piante (U.S. project professional). Jim is a PM practitioner for a multinational company; he delivers presentations on soft skills to project professionals at PM-International Congresses and Events, constantly transmitting his power, positivism, and energy. I have referred many people to Jim. Why? He is a talented, service-oriented person who has gone out of his way to encourage and help me increase the power of networking. Jim has put me in touch with people in his network who are in a position to support me. He distributes his materials at his presentations. Jim is one of those people who keeps giving, giving, and giving. That's why people want to help Jim, which is why his image, visibility, and professionalism are growing more and more internationally.

Another powerful example is my colleague Michel Thiry, who is very active professionally in project management. He has a unique charisma and transmits interest in others for networking purposes. He has increased his professional network very fast in the last few years. How? He observes people at PM Congresses and invites them to talk and join his network. Or he is having dinner together, exchanging experiences, finding ways to do business together, and being proactive. His Valence Network is a real example.

REFERENCES

If you refer someone, make sure that the person mentions your name as the source of the referral. Be explicit. Let's assume you will refer John Smith to your graphic designer, Jane Jones. You might tell John, "Give Jane a call, and please tell her I referred you." Sometimes, you may even call Jane and tell her that John Smith will contact her. Then, remember to ask if John called and how it turned out the next time you see or speak to Jane. You want to reinforce in Jane's mind that you are looking out for her and helping her to grow her business. Be selective. Don't refer to every person you meet. Respect the time of those in your network. Referring to "unqualified" leads will reflect poorly on you. Ask yourself whether or not a particular referral will be valuable to your network partner. Remember that the key is the quality of the leads you supply, not quantity.

COMMUNICATION

Be a good listener. Have you ever been speaking to someone who goes on and on about himself and his business—and never takes a moment to ask about you? We have all run into the "Me, me, me" types—and they are the last people you want to help. So, in your conversations, focus on drawing other people out. Let them talk about their careers and interests. In return, people will perceive you as caring, concerned, and intelligent. You will eventually get your turn to talk about yourself. Call people from time to time just because you care. How do you feel when someone calls you and says, "Hey, I was just thinking about you and was wondering how you are doing?" I'll bet you feel a million bucks. Why don't we make these calls more often if that's the case? Now and then, make it a point to call people in your network to ask how they are doing and offer your support and encouragement. See best practices of communication in Figure 12.2.

That's right. Call just because you care; that is how you want people to treat you. Every December, I pick up the phone and call specific clients I have not spoken with for a long time. Many of these people have not ordered anything from my company in years. My call is upbeat, and my only agenda is to be friendly. I don't try to sell them anything. I appreciate the business they have given me in the past, and I want to hear how they are doing, personally and professionally. Suppose business comes from these calls, that is great. Year after year, I do get business due to making these calls. Someone will say, "I need to order more of these Project management services," or "Our Company is having a sales meeting in six months, and they may want you to do a presentation." Please understand that this is not manipulation or a sales tactic on my part. I am not expecting these people to give me business. I care about how they are doing. Business is simply a by-product of reconnecting with them.

Take advantage of everyday opportunities to meet people. You can make excellent contacts just about anywhere. You never know from what seed your next valuable relationship will sprout. Treat every person as necessary, not just the "influential" ones. Do not be a snob. The person you meet (whether or not they are the boss) may have a friend or relative who can benefit from your product or service. So, when speaking to someone at a

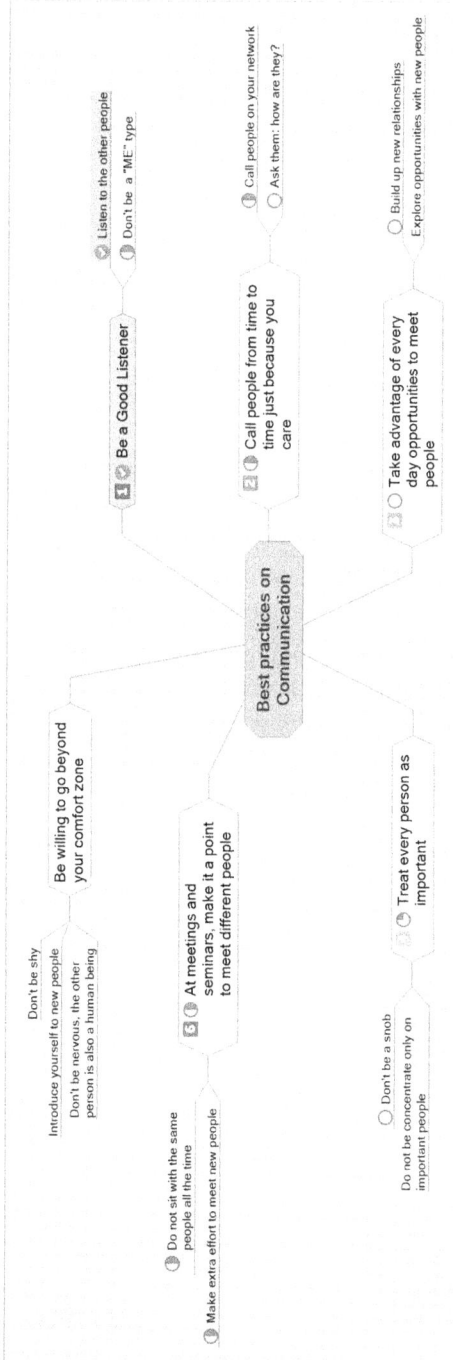

FIGURE 12.2
Best practices of communication.

meeting or party, give that person your undivided attention. And please promise me that you won't gaze around looking for "more important people" to talk to. That bugs me. You are talking with someone, and then he notices someone out of the corner of his eye, someone he deems more important than you. So, he stops listening to you and abruptly breaks away to start a conversation with that other person. Don't do that! Treat every person you encounter with dignity and respect. At meetings and seminars, make it a point to meet different people. Don't sit with the same group at every gathering. While talking with friends for part of the meeting is excellent, you will reap more significant benefits if you try to meet new faces.

In 1996, I was in Washington, DC, to attend a project management training. At lunch, instead of sitting with some friends from HP, I sat at a table where I did not know anyone. Sitting at that table was a man named FA, and we started a conversation. His organization conducts excellent training programs on soft skills for professionals. It turned out that Frank is also a big believer in the importance of attitude. Frank has become a good friend. I am sure glad I did not sit with my friends that day, as I would have missed out on a tremendous opportunity. Be willing to go beyond your comfort zone. For instance, if you want to introduce yourself to someone, Do it! You might hesitate, thinking the person is too significant or busy to speak with you. Even if you are nervous, force yourself to move forward and make contact. You will get more comfortable as time goes on. Ask for what you want. By helping others, you have now earned the right to request assistance yourself. Don't be shy. As long as you have done your best to serve those in your network, they will be more than willing to return the favor.

FOLLOW-UP

Send a prompt note after meeting someone for the first time. Let's say you attend a dinner and make a new contact. Please send a short note as soon as possible explaining how much you enjoyed meeting and talking to those people. Enclose some of your materials and perhaps include information that might interest this person. Ask if there is anything you can do to assist this individual. Be sure to send the note within forty-eight hours after your initial meeting so that it is received while you are still fresh in your contact's mind. Acknowledge powerful presentations or articles. If you hear an exciting presentation or read a great article, send a note to the speaker or writer and tell them how much you enjoyed and learned from their message. One person in a hundred will take the time to do this; be the one who does. I am not saying that speakers and writers often have developed a vast network of people covering a variety of industries, a network you can tap into. When you receive a reference or helpful written materials, always send a thank you note or call to express your appreciation. Follow this suggestion only if you want more references and helpful information. If you don't acknowledge that person sufficiently, they will be much less likely to assist you in the future. Send congratulatory cards and letters. If someone in your network gets a promotion award or celebrates another occasion, write a short note of congratulations. Everyone loves to be recognized, yet very few people take the time to do this. Being thoughtful in this manner can only make you stand out. It is also appropriate to send a card or memorial gift when a family member dies.

BUILDING YOUR NETWORK

The networking suggestions offered above are merely the tip of the iceberg. You should be able to come up with several ideas of your own. How? Go to your library or bookstore, seek out the many excellent books on networking, notice what other people are doing, and adapt their ideas in a way that suits you. Remember that people build networks over time and that significant results usually don't appear immediately. It would help cultivate passion, persistence, and patience to increase your network. Build a solid foundation of relationships and continue to expand and strengthen them. It would help if you put in a lot before you begin reaping the big rewards. Finally, great networking skills are not a substitute for being excellent in your field. You might be terrific, but your efforts will yield disappointing results if you are not talented at what you do and constantly learning and improving. Now, move forward. Select a few of these networking techniques and implement them right away. Get to work serving and improving your network. Then, you will have an army of troops working to help you succeed. Today is a Good Day!

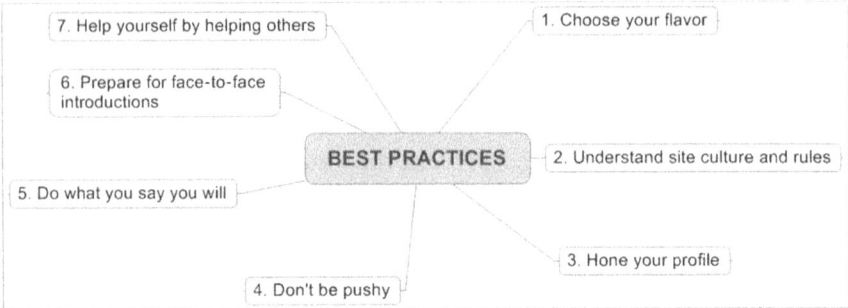

FIGURE 12.3
Best practices.

AVOID NETWORKING GAFFES

I have some best practices to share with you in Figure 12.3.

1. Choose your flavor: Don't jump at every offer to join a professional network.
2. Understand site culture and rules: Before contacting the friend's colleagues who invited you to network, get to know the group's culture.
3. Hone your profile: Review and underline those important characteristics to show others.
4. Don't be pushy: Most professional networks don't like arrogant people.
5. Do what you say you will: Practice authenticity and integrity.
6. Prepare for face-to-face introductions: A face-to-face meeting requires you to respond without the time afforded by e-mail to craft your message. Know what you want for a meeting.
7. Help yourself by helping others: Networking is reciprocal, so do unto others as you'd want done to you. If you can help people, they will be more likely to remember you and return the favor.

CREATE A PERSONAL NETWORKING PLAN

Professional networking is also a project, so you must prepare a plan for that project. You must identify your network contacts, develop a personalized networking plan, and build an administrative process to manage it all. It is essential to ask your network contacts for their help, not for a job. People are delighted to help, but few will have a job to offer you.

1. First-level contacts: These are the hottest prospects and people you know best, current and past colleagues and managers, vendors, consultants, and recruiters with whom you have an established relationship. Your initial contact will likely be via phone, for instance, a quick call announcing you are in the job market and would appreciate advice, assistance, recommendations, or referrals. At the end of each conversation, tell your contacts you'd like to send them a resume to have on file and ask if they prefer mail, fax, or e-mail. Immediately forward your resume with a brief, friendly cover letter, thanking them for any help they can offer and mentioning the positions and industries in which you are interested. If you have not heard back from contacts within three weeks, call and inquire if they have reviewed your resume and if they have any recommendations.

2. Second-level contacts: These are people you know casually. Your initial contact will most likely be fifty percent by phone and fifty percent by mail or e-mail, depending on how comfortable you are in these relationships and how easy it is to connect with each individual. Whenever possible, it is best if the initial contact is a phone call, allowing you to establish a more personal relationship. If you have called a contact, send a resume immediately. If you have not heard back from contacts within three weeks, phone or e-mail them and inquire if they have reviewed your resume and if they have any recommendations.

Once you have developed your list of contacts and determined how to connect with each individual, set up a system to track all your calls, contacts, and follow-up commitments.

CASE STUDY

Since 1993, I have never lost the opportunity to attend international project management congresses yearly. Thanks to that, I know people from Malaysia, Japan, India, the US, Costa Rica, Panamá, Argentina, Peru,

Chile, Mexico, Colombia, Venezuela, Uruguay, Cuba, Brazil, Morocco, Malta, France, Holland, Belgium, Switzerland, Sweden, Norway, Denmark, Russia, Luxemburg, Italia, Greece, Portugal, the UK, Ireland, Arabia, Australia, Rumania, Hungary, Croatia, and Slovenia. When I started my adventure in 1993 to deliver a talk in English, it was challenging because my English level was inferior. Second, it was a big responsibility because I represented my organization internationally and always needed to do my best. And lastly, it was an extraordinary effort on top of that. However, I discovered the internal human being, the power of enthusiasm that encouraged me to move forward and improve my professional skills year by year. I met great people who advised me very positively. I knew people who understood the immense power of networking and the ability to connect with people, sharing experiences, failures, successes, great adventures, and great projects.

I learned that good networking also requires discipline from the professional because you can add professionals to your network, but you must sustain them. It has not been easy for me, but it was not impossible. I maintain my contacts database as alive as possible, have lunch with different colleagues every month, and keep in touch periodically with most of my network colleagues. It has been beneficial for me when I have managed international projects. Getting to know people and having friends worldwide has been very helpful. It is because I take care of my network month by month. I am a member of some project management networks. These communities aim to facilitate the exchange of experiences related to areas of knowledge of "Project Management" to promote personal and professional growth.

SUMMARY

Networking is compelling for you as a project manager; remind some best practices:

- The sooner you start creating a network, the faster you will progress in your career.
- Your success starts with you but grows higher due to your associations and relationships with people.
- Be a good listener.

- Call people from time to time just because you care.
- Treat every person as necessary, not just the "influential roles."
- Send a prompt note after meeting someone for the first time.

TOOL—DEVELOP YOUR NETWORK

I suggest you the following nine activities to develop your professional network. Reflect upon that because you are an excellent professional who is always ready to improve:

1. **Know your goals**.
   The first stage of building your network is figuring out the makeup of the network you want to develop. Consider the outcomes that will be most exciting for you. From there, focus your networking efforts on activities, groups, and people that are most likely to bring you closer to your goals. It can help to shape your network around your long-term career goals.
2. **Acknowledge your value**.
   Unlike mentorship, your network is a reciprocal relationship. In addition to benefiting from your network, there's an expectation that your network will benefit from you in return. Knowing your value can be a helpful confidence booster when building new relationships. As you think about the type of network you want to develop, consider the access, insights, and skills that you feel comfortable offering to members of your network. Additionally, consider which offers may entice the type of people you hope to bring into your network. Improve Your Interpersonal Business Skills—practice and master strategies to improve your professional relationships and help you excel within an organization.
3. **Identify thought leaders**.
   As you conceive of the network you want to build, list the people you consider thought leaders in your field. Include anyone you admire—influencers who provide robust industry analysis, business leaders with career paths you'd like to follow, or individuals currently working in roles you aspire to. While you may not reach out to the people on this list directly, take note of each person's organizations, communities, and affiliations. These may be helpful starting points as you prepare to connect with new and existing contacts.

4. **Consider who you already know.**
   You may already know people who can be valuable additions to your network. Viable candidates for your network can include people you went to school with, have worked with, or have met who work in the same industry as you (or the industry you aspire to work in). All you need to do to transition those relationships to professional ones is to discuss your shared professional interests. If the other person seems receptive, great news: You've just established a professional contact.

Reaching Out to an Acquaintance

If you aren't sure how to reignite a relationship, try something casual and straightforward, like, "Hi, Lucy! I hope you've been well. It's been a minute! Are you still pursuing your cybersecurity certification? I just transitioned into a new role at CyberSecure Industries and would love to catch up. Any interest in grabbing coffee sometime in the next couple of weeks?"

5. **Hone your outreach list.**
   In addition to the people you already know, think about who you want to learn and can reasonably get to know. These would typically be people you have a loose connection with already: colleagues you haven't interacted with much, people you've seen at various industry events, or second-degree contacts (meaning friends or friends). To help you stay organized, list out these potential contacts and include ideas on how you might get in touch with them, whether through direct outreach or by asking a mutual connection to send a letter of introduction.

6. **Identify relevant spaces.**
   The possibilities for growing your network exist far beyond the people you already know and know of. Many people meet new contacts through professional groups, which may come together for social hours, panels, webinars, or other events aimed at career development. As you identify the groups you may want to join, consider the types of spaces you feel most comfortable socializing in. You can likely find active online communities or in-person organizations dedicated to connecting and advancing your industry through an online search or by asking friends and colleagues about groups they engage with. This situation is also a great time to revisit the list of affiliations you noted when you compiled your list of thought leaders. Once you identify the groups you want to join, search their websites and social media profiles for new member information or open events.

7. **Practice your pitch.**

   Practicing how you'll introduce yourself is one way to ease nerves as you prepare to enter new professional environments and meet new people. (Reinforcing to yourself all you have to offer and what you stand to gain is another.) Your delivery may change depending on the person you're contacting and your outreach method—via e-mail, social media, or in person—but the information you share will be consistent. When you meet someone new, be prepared to discuss the following:

   - Who you are and what you do?
   - What you want to learn more about?
   - What you can offer in exchange?
   - How you'd like to move forward?

8. **Say Hi.**

   Saying Hi is the starting point for many relationships—which can require vulnerability. However, fueled by the knowledge of what you want and are prepared to offer, you can start reaching out and building connections. Your relationships will grow over time, but your initial outreach can help set the tone for the relationship you hope to develop. Networking relationships typically fall somewhere between casual friendships and formal work relationships (like the one you might have with your boss or company's CEO, for example). That can be a vast spectrum, and you can use your comfort levels and judgment to gage your most natural balance.

- Casual outreach can be walking up to a stranger at a networking event and simply introducing yourself: "Hi, I'm Jean. This occasion is my first time attending an Emerging Marketing Professionals event. Have you been to any of their panels?"
- Formal outreach might be sending a Slack message to a coworker you'd like to get to know: "Hi, Rohit! I'm on the social media team and enjoyed your presentation on the new app design for the upcoming product launch. I'd love to learn more about your UX design process. We can also discuss how our social media team can help highlight key app features upon launch. Any chance you have time for a 25-minute meeting over the next few weeks?"

9. **Remain engaged.**

   You don't have to become best friends with every person in your network, but to maintain networking relationships, it's essential to invest in them. Some things you can do to stay in touch with your contacts include:

- Exchanging business cards
- Following up after you meet someone new
- Suggesting a time to catch up
- Inviting someone to a networking event you're planning to attend
- Offering introductions between two people in your network
- Making yourself available when someone asks to connect with you

Career paths can sometimes feel unpredictable, and you can't fully know what type of support you may want or need from your network in the future. Staying engaged with your network and remaining open to new possibilities can set you up with a professional support system that you can call upon during any time of need.

13

Conclusions

If you change your attitude, you can adjust your life by managing projects. If I would say that my last five years were a string of successes, I would be lying. I had some failures and some successes too along the way. However, the absolute truth is that I felt happier when I opened my project window and discovered the vast opportunity I had to learn more from all my colleagues and different project stakeholders. Creating a new company was complex, stressful, and challenging but worth it. All the ideas and suggestions discussed in this book are the fruit of my learning as a project manager in organizations, serving my customers, and sharing and learning from colleagues. If I had to explain my great discovery, I would say "project management passion." Apply courage to your projects, control your fears, and try to feel uncomfortable for some time. I am sure you will get something in return.

We all are conscious of the realities of today's economy and the unclear forecasts for the days ahead. However, it would help if you sustained hope as a project manager. Sustaining hope through such times will be a challenging task, even for natural optimists, but believe me, "sustaining hope is possible," and you can do it only if you think you can. Most project managers have a positive attitude some of the time, and it means some project managers have a negative attitude all the time. I found that few project managers have a positive attitude all the time. And you know who the optimistic project managers are. They are evident in the way they act themselves and the way they respond to project issues and problems.

As project managers and leaders, we need to sustain our hope now more than ever. For instance, I have my own business and cannot stop generating new ideas and trying to convert them into new projects.

I always try to maintain my positive attitude, but sometimes I fail, and my motivation decreases. Immediately, I need to go to the toilets and wash my face or walk outside my workplace for a five-minute walk. It would help if you recharged your positive attitude every day. We are human beings; we cannot avoid those failures but recognize and learn from them.

We may choose how we deal with our team members and peers. Be positive and sustain your hope. You will be contagious, and you will be able to infect people with positivism most of the time. I define myself as a "positive mosquito." I am very persistent in all my activities, and mosquitoes are very persistent until they bite you. Do not wait; do it!

BE RESPONSIBLE FOR YOUR ATTITUDE

We cannot control the facts, but we can choose our reactions in front of our problems. So, your work now is to maintain your project window clean enough. My objective with my book is to encourage you to keep your attitude because I assume you always try to have a positive one. My best practice is to set ambitious goals, be prepared, and work hard to achieve those goals, converting our dreams into reality. If you read my previous book, you may have assessed your attitude, but just in case, I propose you review it again and prepare a plan to improve.

The stress generated by the projects we manage, the organizational issues, your colleagues, and your project stakeholders always contaminates you with a negative attitude. You need to be conscious about that and do your attitude check. Always focus on the positive part; repetition is the key. You can do it; you can be positive. Please repeat in your mind: I can be cheerful and think positively regarding any fact that happens to me. Initially, you feel it is difficult, but it is not impossible.

Although we are blessed today, preparing for a better future is desirable. But I discovered that we need to build up our future. Some project managers are waiting for a better future; they say: No problem, we need to pause and be patient, and things will change, but they do not change their mindset. You need to do something new; you need to be creative; you need to use your imagination and propose new ideas. You cannot expect a better future if you cannot change your mindset. Negative or positive thinking is an option for an excellent project manager. Negative thinking is, unfortunately, an automatic process. Positive thinking is a learned self-discipline that you must study and practice daily.

Why does a project manager need to maintain a positive attitude? During the last five years, I had the opportunity to speak about positive attitudes in front of big project management audiences worldwide and obtained different reactions. I need to say that most of them were positive. Still, some project managers and executives asked me several questions regarding the value of maintaining a positive attitude in your projects. Based on several projects I managed and analyzing some business situations I lived in, I can find so many reasons for that:

1. Increase productivity: Most customers prefer positive professionals managing their projects. Having fun increases productivity, but we are not discussing clowning (Englund & Bucero—The Complete Project Manager, 2019).
2. Achieving better performance: Team members perform better with optimistic project managers because their motivation is higher.
3. Positive influence: Project managers with a positive attitude will positively influence their project stakeholders, focusing on problem-solving, not the people.
4. Better results: Management wants positive people leading their projects and inspiring team members to achieve better results.
5. Learning from experiences: A positive attitude provokes "lessons learned" in an easier way, provoking the belief that everybody can be your teacher and helping people to relax and learn.

TAKE CONTROL OF YOUR PROJECTS

Thank you very much for reading this book. I assume you want to increase and develop your professional potential as a project manager or sponsor. This book seeks to be your first step in helping you to live your projects in the way you want to live them. When you are focused on these ideas and thoughts and take action to implement them, you will be on the way to creating some exciting breakthroughs in your projects. I sometimes feel like a UFO (Flying Object Unknown) in the project world in my country

(Spain). It is rare to find an individual who applies the ideas this book suggests daily. And I will repeat it: "It is difficult but not impossible." However, it is scarce for the professional who consistently maintains a positive attitude, knowing that his thoughts will become his reality. You only need to observe your colleague's faces early in the morning and see what happens. It is also rare for a professional to watch their words, knowing they are programming their mind for success, mediocrity, or failure. It is rare for the experienced with the guts to confront their fears because that is where their potential will be developed by doing things she is afraid to do. It is rare for a professional to look for the silver lining in every dark cloud. Finally, finding a professional who commits, follows through with a positive attitude, and has the passion, persistence, and patience to get the job done is rare. I encourage you to be one of those rare professionals. Please join my *Rare's club*. You have the potential to become more than you ever dreamed. You have greatness within you, and your attitude is one of the keys to unlocking that potential. Changing my attitude, I changed many things in my profession. And if a better attitude changed many things in my professional and personal life, it can work for you. I firmly believe we have a daily choice regarding our attitude for that day. We cannot change our past project issues; we cannot change the fact that people will act in a certain way. We cannot modify the inevitable. We can only play on our one string, which is our attitude. I am convinced that life is 10 percent what happens to me and 90 percent how I react to it. We are responsible for our attitude. So then, please take control of your attitude in front of your projects with your team members, sponsors, and other project stakeholders. Move forward and believe in yourself. **Today is a great day!**

MIND MAP OF THIS BOOK

The structure of this book is shown in Figure 13.1 as a mind map.

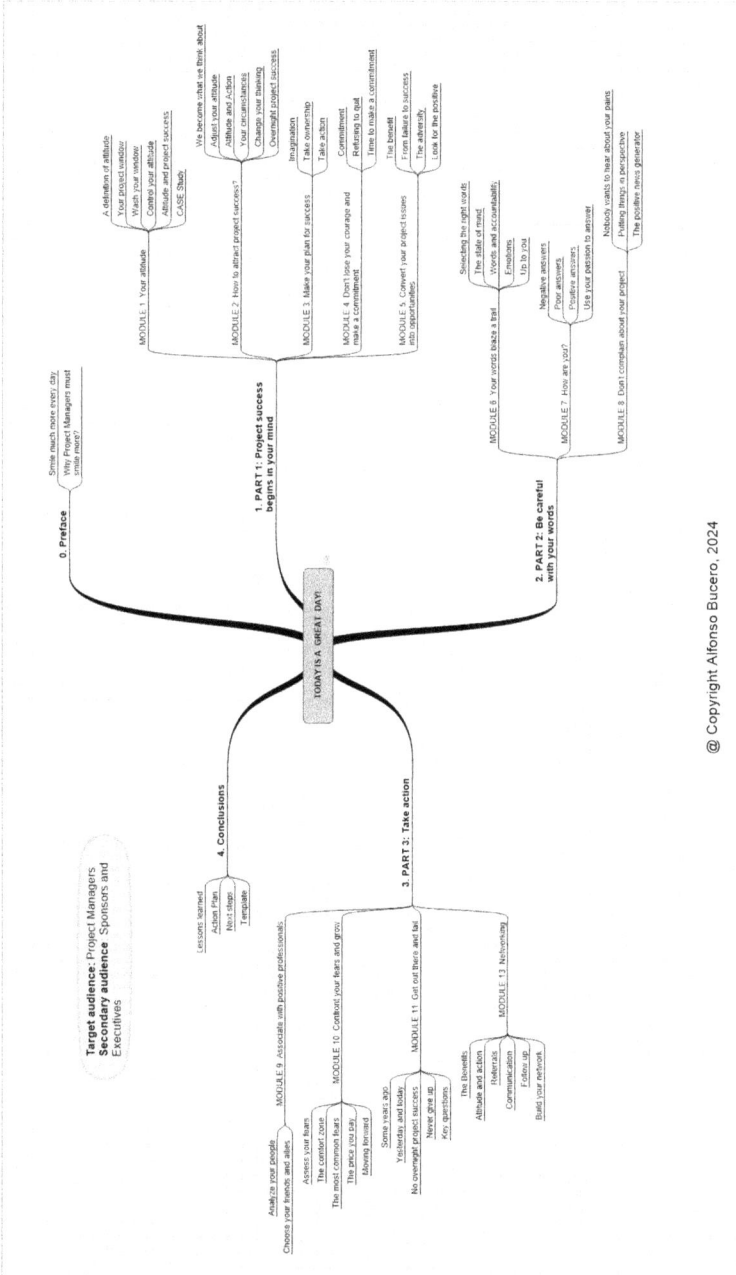

FIGURE 13.1
Today is a Great Day!

Bibliography

Bucero, A. "How to Manage the Change Through Project Management: Red Castle Project—A Real Case." Paper presented at the *World Project Management Week conference*, Cairns, Australia, October 10, 2000.

Bucero, A. "Forging the Future Through PMO Implementation: A Case Study of Sponsorship." Paper presented at the *Project Management Institute's European Project Management Conference*, Cannes, France, June 20, 2002.

Englund, R., & Graham, R. J. (2019). *Creating an environment for successful projects*. Berrett-Koehler Publishers, Oakland, CA.

Englund, R. L. (2000). "Authenticity and integrity." *PM Network, 14*(8), 73–75.

Englund, R. L. "Leading with Power." Paper presented at the *Project Management Institute's Global Congress 2004–North America*, Anaheim, California, October 21–23, 2004.

Englund, R. L. *Environmental Assessment Survey Instrument*. http://home.pacbell.net/muellmar/EASI/EASI%20score+intro.pdf

Englund, R. L., & Bucero, A. (2006). *Project sponsorship: Achieving management commitment for project success*. John Wiley & Sons.

Englund, R. L., Graham, R. J., and Dinsmore, P. C. (2004) *Creating the Project Office: A Manager's Guide to Leading Organizational Change*. San Francisco: Jossey-Bass, 2003.

Frame, J. D. Goleman, D., Boyatzis, R., and McKee, A. *Primal Leadership: Realizing the Power of Emotional Intelligence*. Boston: Harvard Business School Press, 2002.

Graham, R. J., and Englund, R. L. *Creating an Environment for Successful Projects*. (2nd ed.) San Francisco: Jossey-Bass, 2004.

Greenleaf, R. K. *Servant Leadership: A Journey into the Nature of Legitimate Power and Greatness*. Mahwah, N.J.: Paulist Press, 1977.

Harris, P. R., and Moran, R. T. *Managing Cultural Differences*. Houston, Tex.: Gulf, 1996.

Kendall, G. L., and Rollins, S. C. *Advanced Project Portfolio Management and the PMO: Multiplying ROI at Warp Speed*. Fort Lauderdale, Fla.: Ross, 2003.

Kleiner, A. *Who Really Matters: The Core Group Theory of Power, Privilege, and Success*. New York: Currency Doubleday, 2003.

Kouzes, J. M., and Posner, B. Z. *Credibility: How Leaders Gain and Lose It, Why People Demand It*. San Francisco: Jossey-Bass, 1993.

Love, N., and Brant-Love, J. *The Project Sponsor Guide*. Newtown Square, Pa.: Project Management Institute, 2000.

McKenzie, R. *The Relationship-Based Enterprise: Powering Business Success Through Customer Relationship Management*. New York: McGraw-Hill, 2001.

Morris, P.W.G. "Managing the Front End: How Project Managers Shape Business Strategy and Manage Project Definition." Paper presented as the *Project Management Institute's Global Congress*, Edinburgh, Scotland, May 24, 2005.

Pinto, J. K. *Power and Politics in Project Management*. Newtown Square, Pa.: Project Management Institute, 1996.

Project Management Institute. *A Guide to the Project Management Body of Knowledge*. (3rd ed.) Newtown Square, Pa.: Project Management Institute, 2004.

Senge, P. M., and others. *The Fifth Discipline Fieldbook*. New York: Doubleday/Currency, 1994.

Senge, P. M., and others. *The Dance of Change: The Challenges to Sustaining Momentum in Learning Organizations*. New York: Doubleday/Currency, 1999.

Strategic Management Group. *Understanding Project Management*. CD-ROM. Conshohocken, Pa.: Strategic Management Group, n.d.

Thomas, J., Delisle, C. L., and Jugdev, K. *Selling Project Management to Senior Executives: Framing the Moves That Matter*. Newtown Square, PA: Project Management Institute, 2002.

Urban, C. "The Human Factor of Change Management." *Training*, December 22, 2004. http://www.trainingmag.com/training/reports_analysis/feature_display.jsp?vnu_content_id=1000709589

Verma, V. K. *Organizing Projects for Success: The Human Aspects of Project Management*. Newtown Square, PA: Project Management Institute, 1995.

Wattles, W. D. (2022). *The Science of Wallace D. Wattles: Complete Trilogy: The Science of Being Well, The Science of Getting Rich & The Science of Being Great*. DigiCat.

Young, A. *The Manager's Handbook: The Practical Illustrated Guide to Successful Management*. New York: Crown, 1986.

Index

Pages in *italics* refer to figures.

For Product Safety Concerns and Information please contact our EU
representative GPSR@taylorandfrancis.com
Taylor & Francis Verlag GmbH, Kaufingerstraße 24, 80331 München, Germany

www.ingramcontent.com/pod-product-compliance
Lightning Source LLC
Chambersburg PA
CBHW061312220326
41599CB00026B/4848